Quantitative Social Science Research in Practice

Quantitative Social Science Research in Practice: Generating Novel and Parsimonious Explanatory Models for Social Sciences examines quantitative Behavioral Science Research (BSR) by focusing on four key areas:

Developing Novel, Parsimonious, and Actionable Causal Models: Researchers often face challenges in creating new, parsimonious causal models supported by empirical evaluation. A promising approach involves using meta-analytic reviews and more recent studies to identify relevant constructs and hypotheses that would constitute the new causal model.

Exploring the Scope of Context for a Novel Causal Model: The relevance of causal models may vary based on context, such as national or organizational culture, economic and political situations, and feasibility constraints. Behavioural science researchers have struggled to balance rigor and relevance, as theories effective in one context may not be valid in another. This book presents an approach to determine the contextual scope for new causal models.

Guidance to Practice from a Novel Causal Model: Quantitative BSR studies should offer practical guidance, but often this guidance is vague or superficial. This book proposes an approach to highlight actionable insights derived from data analysis of new causal models, ensuring that the research offers concrete guidance for practitioners.

Implementing Guidance from Causal Models: A significant limitation of BSR studies is the lack of clarity on how results can be made actionable for decision-makers, considering the costs and benefits of actions. This book presents a method to make research results actionable, especially for models with high explanatory and predictive power.

The book is designed to be useful for various audiences:

- **Business Managers and Practitioners:** Those conducting or utilizing quantitative BSR for decision-making can find practical approaches for developing and applying causal models.
- **Master's and PhD Students:** Students across disciplines interested in quantitative BSR can gain insight into novel methods for creating strong causal models.
- **Active Scholars:** Researchers aiming to apply new approaches in their work can benefit from the strategies outlined.
- **Professors and Instructors:** Those teaching research methodology or supervising theses can find the book a valuable resource for guiding students in their research projects.

The book aims to advance the field of quantitative BSR by providing robust methodologies for developing, contextualizing, and implementing causal models, ensuring both academic rigor and practical relevance.

Quantitative Social Science Research in Practice

Generating Novel and Parsimonious Explanatory Models for Social Sciences

Charlette Donalds

Kweku-Muata Osei-Bryson

Routledge
Taylor & Francis Group

NEW YORK AND LONDON

Cover: Web Large Image (Public)

First published 2025
by Routledge
605 Third Avenue, New York, NY 10158

and by Routledge
4 Park Square, Milton Park, Abingdon, Oxon, OX14 4RN

Routledge is an imprint of the Taylor & Francis Group, an informa business

© 2025 Taylor & Francis

ISBN: 978-1-032-64819-4 (hbk)
ISBN: 978-1-032-64707-4 (pbk)
ISBN: 978-1-032-67893-1 (ebk)

DOI: 10.1201/9781032678931

Typeset in Minion
by SPi Technologies India Pvt Ltd (Straive)

Contents

Acknowledgements

Deepest gratitude goes to my family, friends, and the co-author for their unwavering support and understanding during the writing process. Their encouragement and dedication kept me motivated and helped me see this project through to completion. Thank you!

Charlette Donalds

To my granddaughters, *Gizelle Kaboré* and *Nima Osei-Caudle*, may your early curiosity of the world in which you find yourselves continue to grow and become as radiant as your enlightening smiles and as deep as you are loved.

Kweku-Muata Osei-Bryson

Preface

PROLOGUE

Information Systems (IS) research can be categorized as belonging to the behavioural science research (*BSR*) category and the design science research (*DSR*) category, with arguably most of the published research belonging in the quantitative *BSR* subcategory. It is this quantitative *BSR* subcategory that is the focus of this book.

Quantitative BSR studies aim to develop an explanatory theory that is in the format of a causal model (e.g. Whetten, 1989) consisting of the **WHATs** (concepts), **HOWs** (links between the concepts), **WHYs** (justification for each link), and the **WHERE** and **WHEN** (i.e. the situations in which the theory applies). With respect to Whetten (1989), context is associated with the *WHERE* and *WHEN*, while the *WHATs* and *HOWs* describe the structure of the causal model. *Quantitative BSR* studies differ in terms of both the *WHATs* and *HOWs*, which are often made clear to the reader, and also the context (*WHO, WHERE,* and *WHEN*), which typically is not made clear to the reader. The concept of relevance should involve the consideration that there may be different 'best' causal models, each of which is valid only for a specific context, and not all contexts.

FOCUS OF THIS BOOK

In this book, we aim to provide support to *quantitative BSR* researchers. Our focus is on the following areas that can reasonably be argued to be in need of attention, particularly for students.

1. **Supporting the Development of Novel, Parsimonious, and Actionable Causable Models:**

   Quantitative BSR studies typically initially involve the development and logical justification of a novel parsimonious and actionable causal model that likely offers a strong explanation for the phenomenon of interest. The traditional approach for the development and justification of such a causal model involves the exploration of existing theories from the knowledge base to generate and justify a set of hypotheses that together constitute the new causal model. A significant obstacle to such exploration of theories is the difficulty that the researcher faces in generating a new parsimonious causal model that is likely to be supported by rigorous empirical evaluation.

According to Osei-Bryson and Ngwenyama (2011), systematic testing of theories (in the format of causal models) and postulating alternative ones are important to advancing the IS discipline, but there is no clearly articulated approach for conducting such an inquiry. In other words, researchers face a significant obstacle in such exploration of theories as they are generally challenged to generate new causal models consisting of novel and a parsimonious set of relevant hypotheses, supported by rigorous evaluation. A potentially promising starting point is to use results from meta-analytic reviews and some more recent studies to identify relevant constructs (*WHATs*) and hypotheses (*HOWs*) that would constitute the new causal model and to use established theories and insights from relevant research to provide justification for each hypothesis (*WHY*).

2. **Exploring the Scope of Context of a Novel Causal Model:**

   The concept of relevance should involve the consideration that there may be different 'best' causal models depending on the *WHERE* including national culture, organizational culture, economic situation (opportunities and constraints), political situation (opportunities and constraints), organizational feasibility constraints (e.g. technological, economic, legal, operational), and other aspects of context. Information Systems (IS) researchers have been challenged to do research that involve both rigor and relevance (e.g. Robey & Markus, 1998). However, as noted by Davison and Martinsons (2016), theories developed and tested in one context might not be adequate or valid for another context. We present an approach that could be used for determining the scope of the context of a new causal model.

3. **Consideration of the Guidance to Practice Offered by a Novel Causal Model:**

   The expectation that quantitative *BSR* studies should involve both rigor and relevance typically require that they offer guidance for practice. Yet often the guidance for practice that appears in published papers is vague and/or superficial, with the reader possibly being left to wonder whether the consideration of such guidance was considered to be an essential part of the research project. We present an approach that could be used to expose actionable guidance that is offered by the results of the data analysis of the new causal model.

4. **Implementation of the Guidance Offered by Causal Model:**

   A major limitation of quantitative BSR studies is that it is often not clear how the relevant results can be made actionable by decision-makers in organizational settings where costs and benefits of actions are of major concern. We present an approach that offers a path to making research results actionable for studies that have causal models with high explanatory and predictive powers.

We expect that this book will be useful to a wide cross section of constituents as follows:

- Business managers and practitioners in disciplines and industries who conduct quantitative behavioural science research projects and/or utilize results from these projects for decision-making purposes.
- Master's- and PhD-level students across varying disciplines who are interested in conducting quantitative behavioural science research.
- Active scholars who are interested in applying new approaches to generate novel, parsimonious, and likely strong causal models.
- College and university professors/instructors who teach research methodology and who supervise Master's and PhD theses.

HIGHLIGHTS OF THIS BOOK

- Given the conceptual formulation of a 'correct' explanatory model is more complex than initially thought (Panovska-Griffiths, Kerr, Waites, & Stuart, 2021), we propose a new process to develop a novel, relevant explanatory model for a given topic of interest, which can be logically justified based on existing theories and insights from previous studies, and which is likely to produce a new strong, parsimonious, empirically validated model.

- Clear guidance is provided to the researcher on how to generate a novel parsimonious model, and an illustrative example of the process is provided.
- Researchers contend that theories developed and tested for one context may not be adequate or valid in another context. Moreover, there is a dearth of focus on the issue of the scope of context since researchers rarely consider context as a lens that may explain research findings/results. We present an approach that could be used for determining the scope of the context of a new causal model.
- In general, variables in a causal model can be characterized as being extrinsic or intrinsic. The options that are available to a decision-maker for making an actionable decision would be different for an intrinsic variable versus an extrinsic variable. Based on our new causal model, we provide an analysis of how considerations of the range of options offered by each of these types of variables (i.e. intrinsic vs. extrinsic) can be used to determine implications for practice.
- For years there has been a *Relevance* vs. *Rigor* debate, and many researchers have attempted to address relevant problems using rigorous methods. Still often a major limitation of such studies is that it is often not clear how the relevant results can be made actionable by decision-makers in organizational settings where costs and benefits of actions are of major concern. We present an approach that offers a path to making research results actionable for studies that have causal models with high explanatory and predictive powers.

CONCLUDING COMMENTS

To the best of our knowledge, the proposed book does not have any known competing text since the proposed book offers several new approaches developed by the authors, which when applied can generate novel, parsimonious causal models that are likely to be strongly supported by empirical analysis, and which can be used for making research results actionable and determining the scope of the context of the generated causal model.

From casual conversation to a compelling book, our collaboration, melding academic rigor with industry know-how, proved immensely rewarding. Each author's unique perspective enriched the discourse, leaving us immensely proud of our work. We hope, whether you're a student, researcher, or practitioner, you'll find the data, analysis, and discussion meaningful in every chapter.

REFERENCES

Davison, R. M., & Martinsons, M. G. (2016). Context is king! Considering particularism in research design and reporting. *Journal of Information Technology, 31*(3), 241–249. https://doi.org/10.1057/jit.2015.19

Osei-Bryson, K.-M., & Ngwenyama, O. K. (2011). Using decision tree modeling to support piercian abduction in is research: A systematic approach for generating and evaluating hypotheses for systematic theory development. *Information Systems Journal, 21*(5), 407–440.

Panovska-Griffiths, J., Kerr, C. C., Waites, W., & Stuart, R. M. (2021). Chapter 10 - Mathematical modeling as a tool for policy decision making: Applications to the COVID-19 pandemic. In A. S. R. Srinivasa Rao & C. R. Rao (Eds.), *Handbook of statistics* (Vol. 44, pp. 291–326): Elsevier.

Robey, D., & Markus, M. L. (1998). Beyond rigor and relevance: producing consumable research about information systems. *Information Resources Management Journal, 11*(1), 7–16. https://doi.org/10.4018/irmj.1998010101

Whetten, D. A. (1989). What Constitutes a Theoretical Contribution? *Academy of Management Review, 14*(4), 490–495. https://doi.org/10.5465/amr.1989.4308371

Authors

Dr. Charlette Donalds is a full-time faculty member of the Mona School of Business & Management (MSBM), The University of the West Indies (UWI), Mona, Jamaica. She holds a PhD in Management Information Systems from The UWI. She has served in several capacities at the MSBM, to include academic director of the Masters in Computer-Based Management Information Systems and Logistics and Supply Chain Management and head for the Decision Sciences and Information Systems Unit. She has authored two books: the first published in 2011 – *Solving Managerial Problems with Spreadsheets and Databases* – and the second more recently in 2022 – *Cybercrime and Cybersecurity in the Global South.*

Dr. Kweku-Muata Osei-Bryson is Professor Emeritus of Information Systems at Virginia Commonwealth University in Richmond, VA. He has been visiting professor of Computing at the University of the West Indies, Mona, and Information Systems at the Ghana Institute of Management and Public Administration. Previously he was professor of Information Systems and Decision Sciences at Howard University in Washington, DC. He has also worked as an Information Systems practitioner in industry and government in the United States and Jamaica. He holds a PhD in Applied Mathematics (Management Science and Information Systems) from the University of Maryland at College Park, a M.S. in Systems Engineering from Howard University, and a B.Sc. in Natural Sciences from the University of the West Indies, Mona, Jamaica.

1

Quantitative Behavioural Science Research: Challenges, Prior, and New Approaches

1.1 INTRODUCTION

Have you ever wondered why you click on certain advertisements, play certain games, choose certain brands, engage in certain seemingly irrational behaviours, and adhere to (or not) information security best practices/policies? The answers lie not just in our conscious choices but also deep within the complexities of our minds. Quantitative behavioural science research (BSR) seeks to explore and illuminate this hidden landscape of our minds by employing numerical data to understand and predict human behaviour.

Quantitative BSR can be described as a dynamic and rigorous field that employs statistical and mathematical methods to investigate and understand human behaviour. It is usually multidisciplinary, drawing on fields such as psychology, sociology, economics, and other related disciplines to examine human behaviour through a quantitative lens. The emphasis on numerical data and statistical analysis distinguishes this research paradigm, and through the application of advanced statistical techniques, researchers can uncover underlying patterns and correlations, providing valuable insights into the complexities of human behaviour.

Through carefully designed experiments, surveys, and observational studies, quantitative behavioural science researchers typically collect and analyse vast amounts of data with the aim to revealing factors that influence or shape our actions and/or behaviours.

DOI: 10.1201/9781032678931-1

1.1.1 Why Is Quantitative BSR Important?

The benefits and applications of quantitative BSR can be vast and transformative. For instance, quantitative BSR can provide objective and generalizable findings applicable to a population and can also offer evidence-based solutions for real-world problems in various domains. Further, by understanding the 'why' behind our behaviours, quantitative BSR can be used to (among others):

  i. *Improve Individual Well-Being*: It can inform interventions to promote healthy habits, increase financial literacy, improve information security compliance behaviour, and foster positive decision-making.
 ii. *Drive Business Success*: It can guide marketing strategies, product design, product mix, and pricing decisions to resonate more effectively with consumers.
iii. *Shape Effective Policy*: It can inform government initiatives to address social challenges and improve public services. It can also inform organizational policies such as business disaster recovery and information security policies.

1.2 QUANTITATIVE BEHAVIOURAL SCIENCE RESEARCH MODEL DEVELOPMENT AND CHALLENGE

Quantitative BSR studies typically initially involve the development and logical justification of a novel parsimonious and actionable causal model that likely offers a strong explanation for a phenomenon of interest and provides a theoretical contribution. According to Whetten (1989), a causal model that provides a theoretical contribution typically contains four essential elements:

  i. **WHAT**s: The factors/constructs/variables that are logically considered to explain the phenomenon of interest. In considering the factors to include in the model, Whetten (1989) advises that *comprehensiveness* (i.e., all the relevant factors are included) and *parsimony* (i.e., what factors are not included because they add little additional value to understanding the phenomenon) are to be contemplated by the scientist.

ii. **HOWs**: The scientist's next step is to consider how the WHATs are related. That is, the links between the factors or, operationally, the *arrows* (links) that connect the *boxes* (factors).

iii. **WHYs**: These are the logical and compelling justifications for including the factors selected and the relationships between them. In other words, the WHYs provide the "theoretical glue that welds the model together" (Whetten, 1989, p. 491).

iv. **WHO, WHEN**, and **WHERE**: These elements constitute the range of the theory. In other words, they provide the boundaries of generalizability of the theoretical model.

It is typical that a quantitative BSR considers/includes the elements as described above, in particular the *WHAT, WHY*, and *HOW*, when developing the study's actionable causal model. What is sometimes less considered is the *WHO, WHEN*, and *WHERE*.

Nonetheless, the development and justification of a quantitative BSR causal model would usually involve the exploration of existing theories from the knowledge base to generate and justify a set of hypotheses that together constitute a new, and perhaps, novel causal model. However, a significant obstacle to such exploration of theories is the difficulty that the scientist faces in generating a new parsimonious causal model that is likely to be supported by rigorous empirical evaluation. To overcome this challenge and to support the researcher in generating new, novel, parsimonious, actionable causal models that are likely to have strong statistical support, we propose that a potentially promising starting point is to use results from *meta-analytic reviews* (MARs) and some more recent studies to identify relevant factors/constructs/variables (*WHATs*) and hypotheses (*HOWs*) that would constitute the new causal model, and to use established theories and insights from relevant literature to provide justification for each hypothesis (*WHY*).

In the quantitative BSR paradigm, MARs are powerful tools for synthesizing and analysing the findings of multiple studies on a phenomenon of interest. MARs offer a comprehensive and statistically robust way to understand the overall evidence presented in multiple studies and draw stronger conclusions than any single study alone (e.g., Cram, D'Arcy, & Proudfoot, 2019; Sommestad, Hallberg, Lundholm, & Bengtsson, 2014). MARs can also aid in addressing the issue of *comprehensiveness* and *parsimony* when determining the WHATs, since they typically identify the factors/constructs/variables that have the strongest statistical significance and effect size on the phenomenon being investigated (see Cram et al., 2019). MARs are also an appropriate potential starting point for our proposed approach as the results of the same can be used to guide future research – by highlighting gaps in the knowledge base or areas of conflicting findings on a specific phenomenon of interest, and by providing valuable directions for further investigation for a scientist (see, e.g., Cram et al., 2019; Sommestad et al., 2014). We contend that combining the results of a MAR and more recent studies will ultimately result in new, novel, and strong hypotheses that will likely be supported by rigorous empirical evaluation.

1.3 PRIOR APPROACHES FOR GENERATING HYPOTHESES

In BSR, several approaches are presented for developing the hypotheses of novel causal models for the domain of interest. For instance, Siponen and Klaavuniemi (2020, p. 2) note that "many IS [information systems] studies report that the hypothetico-deductive (H-D) approach is the most common research approach in IS". Although Siponen and Klaavuniemi (2020) do not fully agree that many IS studies strictly apply a hypothetico-deductive (H-D) method per Hempel or per Popper, other IS researchers over several decades note that IS papers follow an H-D logic and that said method remains the dominant approach used to generate hypotheses and causal models (see, e.g., Hassan, Mathiassen, & Lowry, 2019; Lee, 1991; Orlikowski & Baroudi, 1991; Vessey, Ramesh, & Glass, 2002). Therefore, we will begin our discussion with an overview of the H-D approach, followed by that of other methods, including more recent ones.

1.3.1 Pure Hypothetico-Deductive Approach

The H-D model of theory development (Hempel, 1966; Popper, 1935) can be described as a cyclical process, as shown in Figure 1.1. The scientist first observes some puzzling phenomena in the domain of interest. Based on the scientist observation and existing knowledge, he/she then attempts to produce general statements that he/she believes could explain the phenomena. These statements are the hypotheses, which are testable

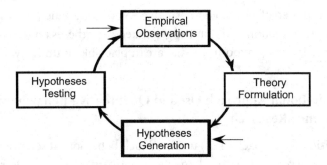

FIGURE 1.1
General Model of Hypothetico-Deductive Logic.

statements, about the relationships between factors/constructs/variables. The scientist then tries to deduce the hypotheses' implications and design experiments to test the logical consequences of the hypotheses. After testing, the scientist analyses the data collected from the experiment to see if it supports the predictions. Based on the analysis, the scientist draws conclusions about the validity of the hypotheses. If the data aligns with the scientist's predictions, the hypotheses gain support, but not proven. On the other hand, if the data contradicts the predictions, the hypotheses are likely falsified and need revision or rejection. Depending on the conclusions, the scientist may refine the hypotheses. For instance, if the hypotheses weren't fully supported but hold promise, the scientist may decide to adjust them based on the findings and conduct further tests. On the other hand, if the hypotheses were clearly falsified, the scientist may decide to abandon them and consider alternative explanations; that is, the hypotheses will be rejected. The H-D method is cyclical because the scientist may revisit previous steps as he/she gathers new evidence and refines his/her understanding of the phenomenon of interest.

While the H-D method provides a powerful framework for conducting quantitative BSR, that is, by systematically formulating hypotheses, deriving predictions, and testing them against evidence, according to Siponen and Klaavuniemi (2020, p. 3), "the H-D method recognizes guessing and imagination in proposing new theories and hypothesis". Thus, the imagination of the scientist/research team could be a limitation of the pure H-D method (per Hempel or per Popper) with respect to the generation of hypotheses. This claim is supported by Dougherty and Hunter (2003), who note that with regard to hypothesis generation, individuals generate only a fraction of the total number of plausible alternative hypotheses when attempting to address a phenomenon and that the generated hypotheses tend to be highly likely, while ignoring hypotheses that are implausible or unlikely.

1.3.2 Traditional Approach Used in Quantitative Behavioural Science Research

The traditional approach (T-A) used in much behavioural science research (BSR) emphasizes a structured and systematic process for understanding and explaining the 'why' behind human behaviours/actions by using data and statistical analysis. This structured and systematic process typically involves the following:

i. *Encountering the Phenomenon*: The scientist/research team is first confronted by a phenomenon for which he/she/team seeks an explanation. This phenomenon becomes the central focus of the investigation.

ii. *Identifying the Gap(s)*: The scientist/research team conducts a thorough literature review of theories and studies related to the observed phenomenon in order to identify the gap(s) in existing knowledge and for which the study has the potential to significantly improve the understanding of the phenomenon.

iii. *Building the Causal Model*: The research team leverages existing knowledge from established theories and prior studies to deduce a causal model. This model depicts the hypothesized relationships between variables; the set of which are the testable hypotheses.

iv. *Rigorous Experimentation*: The hypotheses derived from the causal model are then tested (i.e., to seek confirmation or falsification) through carefully designed experiment(s).

v. *Evaluating and Refinement*: The results from the experiment(s) are analysed statistically to assess whether they support or refute the proposed hypotheses. Supported hypotheses are kept, and later may form the foundation for further exploration (i.e., extended and/or re-tested). Unsupported hypotheses are critically evaluated, potentially leading to modifications of the causal model or the development of new research questions.

vi. *Iteration and Advancement*: The T-A process is rarely linear. As new evidence emerges, the scientist/research team may refine the causal model, test additional hypotheses, or even propose entirely new models. This continual cycle of exploration, testing, and refinement allows for a deeper understanding of the phenomenon and advancement of knowledge in the field.

1.3.3 Abduction Based on Qualitative Data for Behavioural Science Research

According to Timmermans and Tavory (2012, p. 170), "abduction refers to an inferential creative process of producing new hypotheses and theories based on surprising research evidence. A researcher is led away from old to new theoretical insights". Abduction, in the context of qualitative research, refers to a specific type of reasoning used to develop explanations and hypotheses based on qualitative data. It is distinct from deduction and induction, which draws conclusions from existing theories and generalizes from specific observations, respectively. Abduction for BSR based on qualitative data typically involves:

i. *Data Immersion and Analysis*: The scientist/research team carefully analyses qualitative data, such as interviews, observations, or documents on a phenomenon of interest, immersing himself/herself/ themselves in the richness and detail of the information.

ii. *Identifying Patterns and Themes*: Through analysis, the scientist/ research team identifies recurring patterns, themes, and unexpected findings within the data.

iii. *Generating Plausible Explanations*: Based on these patterns and themes, the scientist/research team proposes possible explanations (hypotheses) for the observed phenomena. These explanations are not necessarily based on existing theories but rather emerge from the data itself.

iv. *Evaluating and Refining Explanations*: The scientist assesses the plausibility and coherence of these explanations, considering their fit with the data and existing knowledge. He/she/they may refine, modify, or even discard initial explanations based on further analysis and dialogue.

v. *Seeking Additional Evidence*: To strengthen the explanations, the scientist/research team might seek additional data, either qualitative or quantitative, to confirm or disconfirm the proposed hypotheses. This process is iterative and involves continual refinement and testing of explanations.

This approach is used in mixed-methods research methodologies such as that of Venkatesh, Brown, and Bala (2013) and Niederman and White Baker (2022) and illustrated by Venkatesh, Brown, and Sullivan (2016), who extended the guidelines presented in Venkatesh et al. (2013).

1.3.4 Abduction Based on Quantitative Data for Behavioural Science Research

Recall, abduction is a type of reasoning or inference process that allows the scientist/research team to form explanations or hypotheses for observed data by inferring the most likely cause. In the context of quantitative research, the scientist/research team uses data analytic technique(s) to abduct hypotheses from the data. For instance, Donalds and Osei-Bryson (2017) and Kositanurit, Osei-Bryson, and Ngwenyama (2011) used data mining data analytic technique to abduct new hypotheses about individual information security compliance behaviour and IS end-user performance, respectively. Abduction for BSR based on quantitative data involves:

i. *Leverage Existing Theory*: The scientist/research team uses existing theory to identify highly relevant variables relevant to the phenomenon of interest, but not to identify causal relationships.

ii. *Gather Quantitative Data*: Given the identified relevant variables, the scientist/research team collects quantitative data based on these variables.

iii. *Abduct Hypothesis from Data*: The scientist/research team abducts hypotheses based on the analysis of the quantitative data and/or use data analytic techniques to automatically generate (abducted) hypotheses from the numeric data. Based on the hypotheses, the scientist/research team generates a preliminary model that appears to explain the phenomena of interest.

iv. *Evaluate, Refine, and Provide Justification*: The scientist/research team examines and, if necessary, revises the preliminary model that was previously generated. This revision may be based on the researcher's/research team's knowledge of the domain/theories.

Further, the scientist/research team explores the relevant literature to find logical support and reasoning for each remaining hypothesis.

Examples of this research approach for BSR include the hybrid Peircian abduction framework proposed by Osei-Bryson and Ngwenyama (2011) and the hybrid process for empirically based theory development by Kositanurit et al. (2011), where the latter parallels the T-A for scientific inquiry.

1.4 CONSIDERATIONS FOR A NEW METHOD

In considering whether a new method for generating hypotheses could be of value to the research community, it is necessary to review the strengths and challenges of previously proposed approaches, as discussed in previous sections. Table 1.1 provides an outline of the strengths and challenges of each approach.

Given the strengths and challenges of the considered set of prior approaches identified in Table 1.1, the aim is to present a new approach that at a minimum has the following strengths:

- Generated causal model is novel.
- Generated causal model is parsimonious.
- Generation of hypotheses based on results from a wide range of studies, and implicitly a wide range of theories, such as would be examined in a *MAR*, which by the very nature of a MAR addresses both *comprehensiveness* and *parsimony*.
- Overcome the challenge of reliance on human imagination for hypothesis generation.
- Data collection does not have to be done so as to generate hypotheses.
- Each hypothesis is appropriately logically justified.
- Each hypothesis is likely to be supported in the 'Hypothesis Testing Phase', given the MAR identifies the most influential variables influencing the phenomenon of interest and also the effect size of each variable.

It is likely that our new approach will have its own set of challenges, but the aim is that its challenges should be much less than any of the considered set of prior approaches.

TABLE 1.1

Comparison of Prior Approaches Used in BSR – Strengths and Challenges

Method	Strength(s)	Challenge(s)
Pure H-D	• Generated causal model is novel. • Set of identified hypotheses is not limited by the set of existing theories and the set of prior studies that were considered.	• Generation of hypotheses is limited by reliance on human imagination.
Traditional approach	• Each hypothesis is appropriately logically justified. • Generated causal model is novel.	• Generation of hypotheses is limited by the set of theories and the set of prior studies that were considered.
Abduction – qualitative data	• By not relying solely on existing theories, abduction based on qualitative data can lead to novel and unexpected findings that may have been overlooked by other approaches. • Each hypothesis is appropriately logically justified based on relevant existing theories and prior studies.	• Data collection is done in order to generate hypotheses. • Qualitative analysis, including coding, has to be done appropriately. • Abduction of hypotheses is not automatic. • Generated causal model may not be novel.
Abduction – quantitative data	• Abduction of hypotheses is automatic. • Each hypothesis is appropriately logically justified based on relevant existing theories and prior studies. • Each hypothesis is likely to be supported in the 'Hypothesis Testing Phase'.	• Generated causal model may not be novel. • Limitations in the selected data analytic method(s) used for automatic abduction of hypotheses.

1.5 BOOK'S ORGANIZATION

This book contains seven chapters. This first chapter provides an overview of quantitative BSR research, some methods used for conducting research in this research paradigm, and presents justifications for a new method for generating hypotheses. Chapter 2 provides a detailed description and explanation of our new approach for hypotheses generation and multiple relevant causal models, and in Chapter 3, the proposed approach is illustrated using a case study, which is grounded in the information security policy compliance (ISPC) domain. Chapter 4 then presents the results of an empirical evaluation of the causal model presented in Chapter 3. In Chapter 5, we present a new procedure for more adequately addressing context in the pre-data collection and post-data analysis phases of a given study, that is, given researchers' often inadequate reflection on context before the data collection phase of the study and relatively superficial analysis of context after the data analysis phase of the study. Chapter 6 presents an approach that could be used to expose actionable guidance to practitioners that is offered by the results of a causal model, while also considering feasibility factors. To conclude, in Chapter 7, we discuss how relevant results can be made actionable by decision-makers in organizational settings.

1.6 CONCLUSION

As our world continues to face growing challenges in areas like public health (as evidenced by the COVID-19 pandemic), education, climate, and cyber/information security, quantitative BSR offers valuable tools to understand human behaviours in these and other contexts and to develop effective solutions. By analysing numerical datasets, scientists can identify factors influencing individual and group behaviours in these and other contexts, allowing for tailored interventions and policy decisions.

This book offers a comprehensive exploration of quantitative BSR, highlighting BSR methods and their practical applications. It begins with an overview of the quantitative BSR paradigm, delving into established methods and justifying the need for a new novel quantitative BSR approach. We contend that combining the results of a MAR and at least one more recent study is a good foundation upon which to build this novel approach. The core of the book (Chapters 2 and 3) introduces our new method for generating hypotheses and constructing causal models, which is then demonstrated through a compelling case study in ISPC.

Building on this foundation, Chapter 4 empirically evaluates the presented ISPC causal model, providing insights into its validity. The book then takes a unique perspective by addressing the crucial role of 'context' in BSR studies (Chapter 5). It introduces a procedure for researchers to more thoughtfully integrate 'context' throughout the research process, from pre-data collection analysis to post-analysis interpretation.

Finally, the book focuses on the translation of research findings into real-world impact (Chapter 6). It presents an approach for extracting actionable guidance from causal models, ensuring that results are not only informative but also feasible to implement. In the concluding Chapter 7, we discuss how decision-makers within organizations can leverage these evidence-based insights for transformative action.

Overall, this book is a valuable resource for students, scientists, research teams, and practitioners interested in advancing quantitative BSR. It delves beyond existing methods and provides a framework for developing relevant, robust, novel, parsimonious causal models, rigorously testing them, considering contextual factors, and ensuring that research ultimately drives positive, practical change within organizations or domains of interests.

REFERENCES

Cram, W. A., D'Arcy, J., & Proudfoot, J. G. (2019). Seeing the forest and the trees: A meta-analysis of the antecedents to information security policy compliance. *MIS Quarterly*, 43(2), 525–554. https://doi.org/10.25300/MISQ/2019/15117

Donalds, C., & Osei-Bryson, K.-M. (2017). *Exploring the Impacts of Individual Styles on Security Compliance Behavior: A Preliminary Analysis*. Paper presented at *the SIG ICT in Global Development, 10th Annual Pre-ICIS Workshop*, Seoul, Korea.

Dougherty, M. R. P., & Hunter, J. E. (2003). Hypothesis generation, probability judgment, and individual differences in working memory capacity. *Acta Psychologica, 113*(3), 263–282. https://doi.org/10.1016/S0001-6918(03)00033-7

Hassan, N. R., Mathiassen, L., & Lowry, P. B. (2019). The process of information systems theorizing as a discursive practice. *Journal of Information Technology, 34*(3), 198–220.

Hempel, C. G. (1966). *Philosophy of natural science*. New Jersey: Prentice Hall.

Kositanurit, B., Osei-Bryson, K.-M., & Ngwenyama, O. (2011). Re-examining information systems user performance: Using data mining to identify properties of IS that lead to highest levels of user performance. *Expert Systems with Applications, 38*(6), 7041–7050. https://doi.org/10.1016/j.eswa.2010.12.011

Lee, A. S. (1991). Integrating positivist and interpretive approaches to organizational research. *Organization Science, 2*(4), 342–365.

Niederman, F., & White Baker, E. (2022). The 'case to theory transformation method' for initiating is theory: The process and an illustration using is integration following mergers and acquisitions. *Information Technology & People, 35*(7), 2263–2287. https://doi.org/10.1108/ITP-10-2020-0696

Orlikowski, W. J., & Baroudi, J. J. (1991). Studying information technology in organizations: Research approaches and assumptions. *Information Systems Research, 2*(1), 1–28.

Osei-Bryson, K.-M., & Ngwenyama, O. K. (2011). Using decision tree modeling to support Piercian abduction in IS research: A systematic approach for generating and evaluating hypotheses for systematic theory development. *Information Systems Journal, 21*(5), 407–440.

Popper, K. R. (1935). *The Logic of Scientific Discovery.* Routledge, 513.

Siponen, M., & Klaavuniemi, T. (2020). Why is the hypothetico-deductive (H-D) method in information systems not an H-D method? *Information and Organization, 30*(1), 100287. https://doi.org/10.1016/j.infoandorg.2020.100287

Sommestad, T., Hallberg, J., Lundholm, K., & Bengtsson, J. (2014). Variables influencing information security policy compliance: A systematic review of quantitative studies. *Information Management & Computer Security, 22*(1), 42–75. https://doi.org/10.1108/IMCS-08-2012-0045

Timmermans, S., & Tavory, I. (2012). Theory construction in qualitative research: From grounded theory to abductive analysis. *Sociological Theory, 30*(3), 167–186. https://doi.org/10.1177/0735275112457914

Venkatesh, V., Brown, S. A., & Bala, H. (2013). Bridging the qualitative-quantitative divide: Guidelines for conducting mixed methods research in information systems. *MIS Quarterly, 37*(1), 21–54.

Venkatesh, V., Brown, S. A., & Sullivan, Y. W. (2016). Guidelines for conducting mixed-methods research: An extension and illustration. *Journal of the Association for Information Systems, 17*(7), 435–495. https://doi.org/10.17705/1jais.00433

Vessey, I., Ramesh, V., & Glass, R. L. (2002). Research in information systems: An empirical study of diversity in the discipline and its journals. *Journal of Management Information Systems, 19*(2), 129–174.

Whetten, D. A. (1989). What constitutes a theoretical contribution? *Academy of Management Review, 14*(4), 490–495. https://doi.org/10.5465/amr.1989.4308371

2

A Process for Generating Strong, Novel, and Parsimonious Explanatory Models

2.1 INTRODUCTION

An explanatory theory that is in the format of a causal model (e.g., Whetten, 1989) consists of the *WHATs* (concepts), *HOWs* (links between the concepts), *WHYs* (justification for each link), and the *WHERE* and *WHEN* (i.e., the situations in which the theory applies). Context is associated with the *WHERE*, *WHEN*, and also *WHOs*, while the *WHATs* and HOWs describe the structure of the causal model. Behavioural science research studies differ in terms of both the *WHATs* and *HOWs*, which is often made clear to the reader, and also the context (e.g., *WHO*, *WHERE*, and *WHEN*), which typically is not made clear to the reader. For example, with respect to information security policy compliance (ISPC) studies, a few have focused on the importance of the WHERE dimension of context in terms of its 'national culture' aspect. For example, Chen and Zahedi (2016) in their comparative study on the United States and China argued "*for a context-specific approach to study security behaviours*". The concept of relevance should involve the consideration that there may be different 'best' causal models depending on the *WHERE*, including national culture, organizational culture, economic situation (opportunities and constraints), political situation (opportunities and constraints), and organizational feasibility constraints (e.g., technological, economic, legal, operational). Information systems (IS) researchers have been challenged to do research that involve both rigour and relevance (e.g., Robey & Markus, 1998). However, as noted by Davison and Martinsons (2016), theories developed and tested in one context might not be adequate or valid for another context.

DOI: 10.1201/9781032678931-2

A fundamental tenet of positivist scientific inquiry is the exploration of existing theories to generate and justify a set of hypotheses that together constitute a new causal model. A significant obstacle to such exploration of theories is the difficulty that the researcher faces in generating a new causal model that consists of a novel, actionable, and parsimonious set of relevant hypotheses that are likely to be supported by rigorous evaluation. Such a model may consist of some old and some new hypotheses that taken together provide a strong explanatory model for the phenomenon of interest (e.g., end-user ISPC). However, it should be noted that a model that consists of a set of hypotheses, each of which has been previously proposed, but for which no previous study included that particular set of hypotheses, could also provide a strong and novel explanation for the phenomenon of interest.

According to Osei-Bryson and Ngwenyama (2011), systematic testing of theories (in the format of causal models) and postulating alternative ones is important to advancing the IS discipline, but there is no clearly articulated approach for conducting such an inquiry. In other words, researchers face a significant obstacle in such exploration of theories as they are generally challenged to generate new causal models consisting of novel and a parsimonious set of relevant hypotheses, supported by rigorous evaluation. A potentially promising starting point is to use results from meta-analytic reviews (MARs) and some more recent studies to identify relevant constructs (*WHATs*) and hypotheses (*HOWs*) that would constitute the new causal model, and to use established theories and insights from relevant research to provide justification for each hypothesis (*WHY*).

2.2 PROCESS FOR GENERATING RELEVANT CAUSAL MODELS

Our goal is to develop a relevant, parsimonious causal model that can be logically justified based on established theories and insights from previous studies, and which is likely to produce a new parsimonious and statistically strong, empirically validated model. While Figure 2.1 presents an overview of our new relevant causal model process, the subsections below present a detailed description/explanation of the process. The reader

Stage	Step 0: Initialization	Step 1: Add MAR Hypotheses	Step 2: Add Recent Study Hypotheses	Step 3: Create New & Other Hypotheses	Step 4: Assess Model	Step 5: Create Candidate Models	Step 6: Select Subset of Models (Optional)
Objective	Gather existing knowledge on a phenomenon of interest	Identify MAR causal links	Identify recent study causal links	Identify novel causal links	Assess whether the candidate model can offer potential contribution	Generate multiple candidate models	Select subset of candidate models for empirical evaluation
Activities	• Identify recent meta-analytic review (MAR) • Identify a more recent study • Identify exogeneous variables • Identify control variables	• Identify MAR high-ranked antecedents • Identify MAR empirically validated causal links from literature • Select subset of causal links	• Review results of more recent study • Identify one or more statistically significant causal link not in MAR • Select subset of causal links	• Review existing literature for new causal links • Review existing literature to logically justify causal links • Select subset and integrate causal links	• Assess potential contributions of novel causal links • Add novel links to model to be considered for empirical testing	• Repeat Steps 1–4	• Formulate and solve binary programming problem to select the subset of candidate models to be empirically evaluated
Output	Potential variables	Initial causal model	Extended initial causal model	Integrated causal model	Candidate model	Multiple candidate models	Subset of models to be evaluated

FIGURE 2.1

A New Approach for Generating Relevant Causal Models.

may note that the subsections below present a process that could be used to generate a set of relevant causal models for a given topic of interest. Application of this process assumes that the research team is familiar with the research body of knowledge on the phenomenon of interest (e.g., end-user ISPC), including the set of established theories that have been used in previous studies.

2.2.1 Description of the Process

Step 0: Initialization

a.   Identify the relevant MAR (e.g., Cram, D'Arcy, & Proudfoot, 2019) on the phenomenon of interest (e.g., end-user ISP compliance).

b.   Identify the more recent relevant study on the phenomenon of interest (e.g., Donalds & Osei-Bryson, 2020).

c.   Let N_{Min} an estimate of the number of usable observations that the researcher believes he/she can obtain via sampling the relevant target population. Use a minimum sample size estimation method (e.g., Cohen, 1992; Hair, Ringle, & Sarstedt, 2011; Kock & Hadaya, 2018) to determine for a sample of size N_{Min}, the potential maximum number of arrows pointing at an endogenous variable (a.k.a. construct) in the model. Let this value be denoted as k_{Max} (e.g., 12).

d.   Identify the set of control variables (e.g., *Gender, Age, Experience, Education*, and *Job Level*).

Step 1: Add Existing MAR Derived Hypotheses

a.   Use results from the selected MAR to identify an initial set of highly ranked direct or indirect statistically significant antecedent variables. Let this set be labelled C_{MAR}.

b.   Consult the relevant research literature in order to identify an initial set of empirically validated causal links (say H_{MAR}) each of which involves only variables in C_{MAR}.

c.   Select a subset of the causal links in H_{MAR} which together can be used to form the basis of an initial causal model m_{New1}, such that the maximum number of inbound causal links to any of its endogenous variable of m_{New1} is no more than $(k_{Max} - 2)$.

Step 2: Add Existing Hypotheses Derived from the More Recent Relevant Study

    a. Review the results of the selected more recent relevant study whose set of empirically validated causal links (say H_{Rec}) overlaps with those from the selected MARs (i.e., $H_{MAR} \cap H_{Rec} \neq \varnothing$) but is also different (i.e., $H_{MAR} \neq H_{Rec}$), and which includes at least one statistically significant antecedent variable that was not identified in the MARs (i.e., $C_{MAR} \cap C_{Rec} \neq \varnothing$; $C_{MAR} \neq C_{Rec}$).

    b. From these results, select a subset of the causal links in H_{Rec} but not in H_{MAR} that can be integrated with the initial causal model m_{New1} to form the model m_{New2}, such that the maximum number of inbound causal links to any endogenous variable of m_{New2} is no more than $(k_{Max} - 1)$. Let its set of variables of m_{New2} be C_{New2}.

Step 3: Create New and Other Hypotheses

    a. Given m_{New2}, review established existing theories and previous studies relevant to the topic of interest, to determine if any new and/or other interesting causal links involving the variables in C_{New2} can be logically justified using such theories and studies.

    b. From these results, select a subset of these new causal links that can be integrated with the model m_{New2} to form a new model m_{New3}, such that the maximum number of inbound causal links to any endogenous variable of m_{New3} is no more than k_{Max}. Selection preference should be given to causal links that have not been previously evaluated. Let the set of causal links of m_{New3} be labelled H_{New3}.

Step 4: Assess the Potential of the Current Candidate Model

    a. The research team should assess whether evaluation of this m_{New3} has the potential to offer valuable contributions to the literature. An important consideration here is that m_{New3} should be novel (i.e., H_{New3} is not a subset of the hypotheses of any model that had been addressed in any single previous study).

    b. If the research team determines that it has such potential, then m_{New3} should be added to M, the current set of novel candidate causal models that are to be considered for empirical evaluation.

Step 5: Generate Multiple Candidate Models

    a. Steps 1–4 can be repeated until the research team is sufficiently satisfied with the set of generated models.

Step 6: Automatically Select Models for the Empirical Evaluation (Optional)

   a.   Formulate and solve the binary integer programming problem **Select the Candidate Models (SCM)** (see Appendix A).

2.2.2 Explanation and Justification of the Process

In developing and evaluating a novel, parsimonious explanatory model, the research team faces at least five concerns:

i. **Novelty of the Model:** Step 3 aims to achieve this with some assistance from Steps 1 and 2; Step 4 assesses whether novelty has been achieved.

ii. **Parsimony:** In order to be parsimonious, the model should not include too many variables and causal links, even though there could be many variables and links that could reasonably be considered for inclusion in the model. One approach to addressing this concern is to develop and evaluate multiple rigorously developed novel parsimonious causal models. Step 5 (i.e., repeat Steps 1–4) would achieve the development of multiple novel parsimonious causal models.

iii. **Adequate Sample Size:** Evaluation of the model typically involves sampling, and in many cases, the size of the resulting set of usable observations limits the maximum number inbound causal links to any endogenous variable. Substep 0-c generates k_{Max}, which is then used in Steps 1–3 with the aim of ensuring that this concern is addressed.

iv. **Empirical Evaluation Success:** The models developed by executing Steps 0–4 would have been rigorously developed including having logical and/or previous empirical support by the research literature. While there is no guarantee that given causal links would be empirically supported in the research team's study, even the lack of support offers the opportunity to offer insights on conditions under which such support would hold. Further, given that Steps 5 and 6 would result in the generation and evaluation of multiple novel, parsimonious models, then there is a greater likelihood of achieving empirical evaluation success than if only a single model was evaluated.

v. **Complete-able Questionnaire:** The number of items on the questionnaire can affect the number of received usable observations. Step 6 aims to address this concern while allowing for the evaluation of multiple novel parsimonious models.

2.3 CONCLUSION

In this chapter, we presented and logically justified our new process for generating relevant, strong, novel, and parsimonious explanatory models. This process offers several benefits including:

- Selection of highly ranked, previously empirically validated antecedents of the variable of interest.
- Each generated causal model is novel and likely to offer strong explanatory power with respect to the given dependent variable.
- Generation of hypotheses based on results from a wide range of studies, and implicitly a wide range of theories, such as would be examined in a MAR.
- Overcome the challenge of reliance on human imagination for hypotheses generation.
- Data collection does not have to be done in order to generate hypotheses.
- Each hypothesis is appropriately logically justified.
- Each hypothesis is likely to be supported in the 'Hypotheses Testing Phase', given that each predictor variable had strong empirical support in the results of the selected MAR and the selected relevant recent study.

REFERENCES

Chen, Y., & Zahedi, F. M. (2016). Individuals' internet security perceptions and behaviors: Polycontextual contrasts between the United States and China. *MIS Quarterly, 40*(1), 205–222.

Cohen, J. (1992). Quantitative methods in psychology: A power primer. *Psychological Bulletin, 112*(1), 155–159.

Cram, W. A., D'Arcy, J., & Proudfoot, J. G. (2019). Seeing the forest and the trees: A meta-analysis of the antecedents to information security policy compliance. *MIS Quarterly, 43*(2), 525–554. https://doi.org/10.25300/MISQ/2019/15117

Davison, R. M., & Martinsons, M. G. (2016). Context is king! Considering particularism in research design and reporting. *Journal of Information Technology, 31*(3), 241–249. https://doi.org/10.1057/jit.2015.19

Donalds, C., & Osei-Bryson, K.-M. (2020). Cybersecurity compliance behavior: Exploring the influences of individual decision style and other antecedents. *International Journal of Information Management, 51*. https://doi.org/10.1016/j.ijinfomgt.2019.102056

Hair, J. F., Ringle, C., & Sarstedt, M. (2011). PLS-SEM: Indeed a silver bullet. *Journal of Marketing Theory and Practice, 9*(2), 139–151.

Kock, N., & Hadaya, P. (2018). Minimum sample size estimation in PLS-SEM: The inverse square root and gamma-exponential methods. *Information Systems Journal, 28*(1), 227–261. https://doi.org/10.1111/isj.12131

Osei-Bryson, K.-M., & Ngwenyama, O. K. (2011). Using decision tree modeling to support Piercian abduction in IS research: A systematic approach for generating and evaluating hypotheses for systematic theory development. *Information Systems Journal, 21*(5), 407–440.

Robey, D., & Markus, M. L. (1998). Beyond rigor and relevance: Producing consumable research about information systems. *Information Resources Management Journal, 11*(1), 7–16. https://doi.org/10.4018/irmj.1998010101

Whetten, D. A. (1989). What constitutes a theoretical contribution? *Academy of Management Review, 14*(4), 490–495. https://doi.org/10.5465/amr.1989.4308371

3

Illustration of the Process for Generating Strong, Novel, and Parsimonious Causal Models

3.1 INTRODUCTION

For our illustrative example, our topic of interest is employees' information security policy compliance (*ISPC*). For our relevant meta-analytic review (*MAR*), we used the relatively recent study of Cram et al. (2019) from which we selected predictors with the high explanatory power of employees' security policy compliance intentions and behaviours (see Table 3.1). For our more recent ISPC study, we use that of Donalds & Osei-Bryson (2020). To provide logical justification for our set of new and previously proposed hypotheses, we draw upon the established theories/literatures of theory of planned behaviour (TPB), protection motivation theory (PMT), deterrence theory (DT), information security awareness (ISA), general security orientation (GSOR), and personal norms (PN). In Section 3.3, we provide the logical justification for each hypothesis, and in Chapter 4, we present the empirical evaluation of the model (Figure 3.1).

3.2 OVERVIEW OF THE CASE STUDY

Employees' noncompliance with information systems (IS) security policies/ guidelines constitutes a major insider threat for their organisations. In fact, a recent industry report reveals alarming statistics relating to insider threat incidents. The Ponemon Institute (2023) report reveals that from 2018 to 2023: (i) the average cost of insider threats skyrocketed, rising some 95% to

DOI: 10.1201/9781032678931-3

TABLE 3.1

Highly Ranked Predictors of ISPC Based on MAR Study

Category/Variable	Raw Relative Weight	Percentage of R^2
Personal Norms & Ethics	0.106	20.66
Attitude	0.091	17.73
Normative Beliefs	0.069	13.38
Self-Efficacy	0.039	7.52
Response Cost	0.037	7.28
SETA	0.032	6.27
Response Efficacy	0.032	6.26
Punishment Severity	0.028	5.50
Threat Severity	0.027	5.24
Detection Certainty	0.024	4.61
Punishment Expectancy	0.012	2.27
Rewards	0.011	2.10
Resource Vulnerability	0.006	1.18
Total R^2	**0.514**	

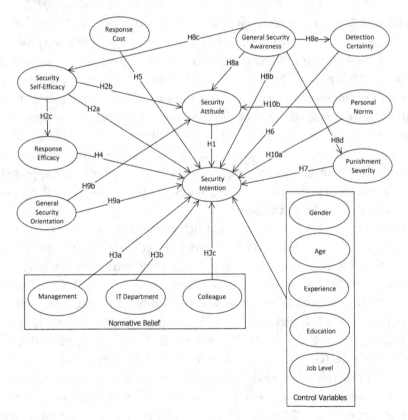

FIGURE 3.1

Information Security Policy Compliance Model.

$16.2 million; (ii) 55% of reported insider threat incidents were as a result of negligent employees or contractors, with an average annual cost of $7.2 million to remediate these incidents; (iii) criminal or malicious insiders accounted for 25% of the reported incidents with an average cost of $701,500 per incident, while credential theft (i.e., outsmarted employees) represented 20% of reported incidents with an average cost per case of $679,621; (iv) malicious insiders were most likely to email sensitive data to outside parties (67%); (v) 71% of organizations surveyed suffered 21–40 incidents per year; and (vi) it takes 86 days on average to detect and contain an insider threat incident. Given the findings that insiders account for a large percentage of security incidents and the significant impacts these incidents have on organizations, a reasonable approach to safeguard organisations' IS assets is for them to focus on employees' intentions and behaviours.

To cope with increased threats, organisations have implemented not only technological solutions such as intrusion detection systems, anti-malware, and antivirus applications, but also IS security policies/guidelines (ISPs) to regulate employees IS security-related behaviours. ISPs are defined as "*a set of formalized procedures, guidelines, roles and responsibilities to which employees are required to adhere to safeguard and use properly the information and technology resources of their organizations*" (Lowry & Moody, 2015, p. 434). However, the literature suggests that ISPs are not viewed by employees as hard and fast rules to follow; instead, employees often ignore them, circumvent them, and/or do not readily comply with the ISP's prescribed behaviours (Ifinedo, 2012; Lowry & Moody, 2015; Vance, Siponen, & Pahnila, 2012). Thus, additional studies to advance our knowledge about factors that may improve ISP compliance are appropriate. Our work here does just that and is designed to complement the growing body of knowledge in the area.

Over the past decade, a diverse and rich plethora of research postulating diverse variables and relationships have been considered that may drive, inhibit, or increase employees' compliance intention with ISPs (see, e.g., Cram et al. (2019) recent MAR of the literature in this area). Collectively, research in this area is rooted in widely used behavioural, criminology, and psychology theories such as TPB (Ajzen, 1991), theory of reasoned action (Ajzen & Fishbein, 1980; Fishbein & Ajzen, 1975), PMT(Rogers, 1975, 1983), DT(Gibbs, 1975), and social bond theory (SBT; Hirschi, 1969). While improving our knowledge and understanding of employees' security compliance (or non-compliance) intentions and behaviours, according to Cram et al. (2019, p. 526), "*the number of competing theoretical* perspectives and inconsistencies in the reported findings have yielded certain unresolved conflicts". Cram et al. (2019) further argued that not only is there a lack of consensus regarding the key drivers of security policy compliance but also that researchers have set about finding the best individual predictors of security policy compliance in a piecemeal fashion.

Given this diverse set of well-reasoned and rigorously empirically validated models with different sets of direct and indirect statistically significant determinants (antecedents) of ISP compliance intention and behaviours, one perspective to such a situation is to view these models as competitors, that is, which one is the 'best'; another perspective is to view them as offering descriptions of legitimate alternate paths to achieving end user IS security compliance, some of which are highly relevant for a given context.

3.3 APPLICATION OF THE CAUSAL MODEL GENERATING PROCEDURE

Here we present an application of *Steps 0–4* of the Causal Model Generating Procedure to our Case Study example (see Chapter 2). The reader may recall that *Step 5* involves repeating *Steps 1–4* until the scientist/research team is sufficiently satisfied with the set of generated relevant models.

3.3.1 Step 0: Initialization

Sub-Step 0-a: Identify the relevant MAR on the phenomenon of end-user *ISPC*.
Output: (e.g., Cram et al., 2019).
Sub-Step 0-b: Identify the more recent relevant study on the phenomenon of interest.
Output: (e.g., Donalds & Osei-Bryson, 2020).
Sub-Step 0-c: Let N_{Min}, an estimate of the number of usable observations that the scientist/research team believes he/she/they can obtain via sampling the relevant target population. Use power analysis to determine for a sample of size N_{Min}, the potential maximum number of arrows pointing at an endogenous variable (a.k.a. construct) in the model. Let this value be denoted as k_{Max}.
Output: $k_{Max} = 1$.

Sub-Step 0-d: Identify the set of control variables.
Output: *Gender, Experience, Tenure, Education, Job Level.*

3.3.2 Step 1: Add Existing Meta-Analytic Review Derived Hypotheses

Sub-Step 1-a: Use results from the selected MAR to identify an initial set of highly ranked direct or indirect statistically significant anteced-ent variables. Let this set be labelled C_{MAR}.
Input: Table 3.1 from Cram et al. (2019).

We chose to omit *Rewards* (since some organizations do not offer *Rewards* for ISPC) and *Threat Severity* (based on its relative low weighting of the PMT constructs), and *SETA* (as it likely to be correlated with *General Security Awareness* of the selected recent study).

Output: $C_{MAR} =$

{*Personal Norms & Ethics, Attitude, Normative Beliefs, Self-Efficacy, Response Cost, Response Efficacy, Punishment Severity, Threat Severity, Detection Certainty*}

Sub-Step 1-b: Consult the relevant research literature in order to iden-tify an initial set of empirically validated causal links (say H_{MAR}) each of which involves only variables in C_{MAR}.

C_{MAR}: Selected Variables	H_{MAR}: Initial Set of Validated Causal Links
• *Personal Norms & Ethics*	IS security *Attitude* → ISP *Compliance Intention*
• *Attitude*	IS security *Self-Efficacy* → ISP *Compliance Intention*
• *Normative Beliefs*	IS security *Self-Efficacy* → IS security *Attitude*
• *Self-Efficacy*	IS security Self-Efficacy → *Response Efficacy*
• *Response Cost*	*Response Efficacy* → ISP *Compliance Intention*
• *Response Efficacy*	*Response Cost* → ISP *Compliance Intention*
• *Punishment Severity*	*Punishment Severity* → ISP *Compliance Intention*
• *Detection Certainty*	*Detection Certainty* → ISP *Compliance Intention*
	Personal Norms → IS security *Attitude*
	Personal Norms → ISP *Compliance Intention*

Sub-Step 1-c: Select a subset of the causal links in H_{MAR} which together can be used to form the basis of an initial causal model m_{New1}, such that the maximum number of inbound causal links to any of the endogenous variable of m_{New1} is no more than $(k_{Max} - 2)$.

Output: All of the causal links in H_{MAR} were selected, so the initial causal model m_{New1} consists of all of the causal links outputted in Sub-Step 1-b.

3.3.3 Step 2: Add Existing Hypotheses Derived from the More Recent Relevant Study

Sub-Step 2-a: Review the results of the selected more recent relevant study whose set of empirically validated causal links (say H_{Rec}) overlaps with those from the selected MARs (i.e., $H_{MAR} \cap H_{Rec} \neq \emptyset$) but is also different (i.e., $H_{MAR} \neq H_{Rec}$), and which includes at least one statistically significant antecedent variable that was not identified in the MARs (i.e., $C_{MAR} \cap C_{Rec} \neq \emptyset$; $C_{MAR} \neq C_{Rec}$).

H_{Rec}: Validated Causal Links from Recent Study	C_{Rec}: Associated Variables
Dominant Decision Style → IS security *Self-Efficacy*	• Dominant Decision Style
Dominant Decision Style → ISP *Compliance Intention*	• Dominant Decision Orientation
Dominant Decision Orientation → General Security Orientation	• General Security Orientation
IS security *Self-Efficacy* → ISP *Compliance Intention*	• General Security Awareness
General Security Awareness → ISP *Compliance Intention*	• ISP *Compliance Intention*
General Security Orientation → ISP *Compliance Intention*	• IS security *Self-Efficacy*
$C_{MAR} \cap C_{Rec}$ = {ISP *Compliance Intention*, IS security *Self-Efficacy*} $\neq \emptyset$	

Sub-Step 2-b: From these results, select a subset of the causal links in H_{Rec} but not in H_{MAR} that can be integrated with the initial causal model m_{New1} to form the model m_{New2}, such that the maximum number of inbound causal links to any endogenous variable of m_{New2} is no more than $(k_{Max} - 1)$. Let its set of variables of m_{New2} be C_{New2}.

Selected Causal Links from ($H_{Rec} - H_{MAR}$)	Associated Variables
General Security Awareness → ISP *Compliance Intention*	• General Security Awareness
General Security Orientation → ISP *Compliance Intention*	• General Security Orientation

C_{New2}: **Associated Causal Links of Model** m_{New2}	**Associated Variables**
IS security *Attitude* → ISP *Compliance Intention*	• *General Security Awareness*
IS security *Self-Efficacy* → ISP *Compliance Intention*	• *General Security Orientation*
IS security *Self-Efficacy* → IS security *Attitude*	• *Personal Norms & Ethics*
IS security Self-Efficacy → *Response Efficacy*	• *Attitude*
Response Efficacy → ISP *Compliance Intention*	• *Normative Beliefs*
Response Cost → ISP *Compliance Intention*	• *Self-Efficacy*
Punishment Severity → ISP *Compliance Intention*	• *Response Cost*
Detection Certainty → ISP *Compliance Intention*	• *Response Efficacy*
Personal Norms → IS security *Attitude*	• *Punishment Severity*
Personal Norms → ISP *Compliance Intention*	• *Detection Certainty*
General Security Awareness → ISP *Compliance Intention*	
General Security Orientation → ISP *Compliance Intention*	

3.3.4 Step 3: Create New and Other Hypotheses

Sub-Step 3-a: Given m_{New2}, review established existing theories and previous studies relevant to the topic of interest, to determine if any new and/or other interesting causal links involving the variables in C_{New2} that can be logically justified using such theories and studies. We provide a review of relevant theories below, and Table 3.2 provides the selected constructs, their definitions, and the relevant theories/literatures with which they are associated.

Label	New Interesting Causal Links Involving Variables of C_{New2}
H3a	*Norm. Belief* (Management) → ISP *Compliance Intention*
H3b:	*Norm. Belief* (IT-Dept.) → ISP *Compliance Intention*
H3c:	*Norm. Belief* (Colleague) → ISP *Compliance Intention*
H8a:	*General Security Awareness* → IS security *Attitude*
H8c:	*General Security Awareness* → IS security *Self-Efficacy*
H8d:	*General Security Awareness* → *Punishment Severity*
H8e:	*General Security Awareness* → *Detection Certainty*
H9b:	*General Security Orientation* → IS security *Attitude*

TABLE 3.2

Selected Constructs and Their Associated Theories/Literatures

Construct	Theory/ Literature	Definition
Security Compliance Attitude (SATT)	TPB	The degree to which the performance of the ISP compliance behaviour is positively valued by the employee (Bulgurcu et al., 2010).
Normative Belief (BELF)		An employee's perceived social pressure about compliance with the requirements of the ISP caused by behavioural expectations of such important referents as executives, colleagues, and managers (Bulgurcu et al., 2010).
IS Policy Compliance Intention (SINT)	TPB; PMT	An employee's intention to protect the information and technology resources of the organisation from potential security breaches by complying with the requirements of the ISP (Bulgurcu et al., 2010).
Security Self-Efficacy (SELF)		An employee's ability or judgement of his or her ability to perform information security actions in such a manner that minimizes the risk of security breaches/incidents (Bandura, 1977; Donalds & Osei-Bryson, 2017).
Response Cost (COST)	PMT	The employee's belief about the perceived opportunity costs in terms of monetary, time, and effort expended in complying with the ISP (Ifinedo, 2012).
Response Efficacy (RESP)		The employee's perception that complying with the ISP will help in removing or preventing threats to IS assets (Ifinedo, 2012; Siponen et al., 2014).
Punishment Severity (PUNS)	DT	The harshness of punishment associated with committing an act of ISP noncompliance (Cheng, Ying, Li, Holm, & Zhai, 2013; Johnston et al., 2015).
Detection Certainty (DETC)		The employee's perception of the likelihood of being caught for an act of ISP noncompliance (D'Arcy et al., 2009; Herath & Rao, 2009b; Li, Zhang, & Sarathy, 2010).
General Security Awareness (GSAW)	ISA	The employee's overall awareness (mindfulness), understanding, and knowledge of potential issues related to IS security and their ramifications (Bulgurcu et al., 2010; Donalds & Osei-Bryson, 2017).
General Security Orientation (GSOR)	HBM	The employee's predisposition and interest concerning practicing IS security actions (Donalds & Osei-Bryson, 2017; Ng et al., 2009).
Personal Norms (PNRM)	SBT	The employee's own values and views regarding ISP compliance (Ifinedo, 2014).

Sub-Step 3-b: From these results, select a subset of these new causal links that can be integrated with the model m_{New2} to form a new model m_{New3}, such that the maximum number of inbound causal links to any endogenous variable of m_{New3} is no more than k_{Max}. Selection preference should be given to causal links that have not been previously evaluated. Let the set of causal links of m_{New3} be labelled H_{New3}.

Label	H_{New3}: **Causal Links of Model** m_{New3}
H1:	IS security *Attitude* → ISP *Compliance Intention*
H2a:	IS security *Self-Efficacy* → ISP *Compliance Intention*
H2b:	IS security *Self-Efficacy* → IS security *Attitude*
H2c:	IS security Self-Efficacy → *Response Efficacy*
H3a:	*Norm. Belief* (Management) → ISP *Compliance Intention*
H3b:	*Norm. Belief* (IT-Dept.) → ISP *Compliance Intention*
H3c:	*Norm. Belief* (Colleague) → ISP *Compliance Intention*
H4:	*Response Efficacy* → ISP *Compliance Intention*
H5:	*Response Cost* → ISP *Compliance Intention*
H6:	*Punishment Severity* → ISP *Compliance Intention*
H7:	*Detection Certainty* → ISP *Compliance Intention*
H8a:	*General Security Awareness* → IS security *Attitude*
H8b:	*General Security Awareness* → ISP *Compliance Intention*
H8c:	*General Security Awareness* → IS security *Self-Efficacy*
H8d:	*General Security Awareness* → *Punishment Severity*
H8e:	*General Security Awareness* → *Detection Certainty*
H9a:	*General Security Orientation* → ISP *Compliance Intention*
H9b:	*General Security Orientation* → IS security *Attitude*
H10a:	*Personal Norms* → IS security *Attitude*
H10b:	*Personal Norms* → ISP *Compliance Intention*

3.3.5 Review of the Relevant Theories/Literature Based on the New Causal Model Antecedents

3.3.5.1 Theory of Planned Behaviour

Resulting from the *Steps 1–3*, several antecedents have been selected to be included in our new robust, novel, and parsimonious causal model, which are related to the TPB (Ajzen, 1991); hence, a brief review of the theory is now presented.

The TPB (Ajzen, 1991) has been applied across many different contexts to explain why individuals behave the way they do and what drive their behaviour. In general, the TPB postulates that *attitude, subjective norm* (a.k.a. *normative belief*), and *perceived behavioural control* – comprised of *self-efficacy* and *controllability* – together shape an individual's *intentions* and *behaviours*. A core argument of the TPB is that the stronger the intention to engage in a behaviour, the greater the likelihood that the actual behaviour will be executed. The role of intention as a predictor of behaviour, for example, ISPC intention (ISPCI), is well established in the IS literature (e.g., Bulgurcu, Cavusoglu, & Benbasat, 2010; Herath & Rao, 2009b; Moody, Siponen, & Pahnila, 2018; Rajab & Eydgahi, 2019).

Although IS scholars argue that the TPB exclusively is insufficient in explaining a significant amount of variance in ISPC behaviour/intention (e.g., Sommestad et al., 2015), numerous empirical studies reveal strong support for the constructs of the TPB as significant predictors of ISPC intention (ISPCI). Specifically, attitude is found to be a strong antecedent of ISPCI (e.g., Bauer & Bernroider, 2017; D'Arcy & Lowry, 2019; Moody et al., 2018) as well as subjective norm (e.g., Moody et al., 2018; Yazdanmehr & Wang, 2016), normative belief (e.g., Bulgurcu et al., 2010; Siponen, Mahmood, & Pahnila, 2014), and self-efficacy (e.g., D'Arcy & Lowry, 2019; Moody et al., 2018). Furthermore, Cram et al. (2019) identify the TPB constructs as being some of the highest relative important antecedents of ISPCI (see Table 3.1). Resulting from *Steps 1–3*, we selected the TPB-related antecedents: *attitude, normative belief,* and *self-efficacy.*

3.3.5.2 Protection Motivation Theory

PMT is another relevant theory from which antecedents of ISPC, resulting from *Steps 1–3*, were selected to be included in our new causal model. Therefore, we now present a brief discussion of the theory.

PMT (Rogers, 1975) explains fear appeals and how people cope with them. Generally, PMT helps us understand individuals' behaviour/actions based on threats posed to themselves and/or their surroundings. The PMT describes coping with a threat as the result of two coping processes, a threat appraisal process and a coping appraisal process. In the ISPC context, threat appraisal is the employees' assessment of the severity of information security (InfoSec) threat, while coping appraisal is the ability of the employee to comply with the ISP and whether such compliance is effective in reducing the InfoSec threat (Donalds & Barclay, 2022).

While empirical results generally support PMT in the context of explaining employees' willingness to comply with ISPs (e.g., Johnston, Warkentin,

& Siponen, 2015; Rajab & Eydgahi, 2019), there are divergence and inconsistencies in the findings. Specifically, in terms of the circumstances under which certain PMT-based relationships hold as well as whether the full nomology of PMT constructs is essential. Demonstrably, self-efficacy has no statistically significant impact on university employees' ISPCI (Rajab & Eydgahi, 2019), yet self-efficacy is shown to increase ISPCI in other organizational settings (e.g., Johnston et al., 2015; Siponen et al., 2014).

Resulting from applying *Steps 1–3* of our causal model generating procedure, we included *self-efficacy, response cost,* and *response efficacy* in our new causal model. Given our objective of including some of the variables from the selected recent study that were not included in the MAR, then to have a parsimonious model, it would be necessary to exclude some of the predictors identified in the MAR (see Table 3.1). We chose to omit *Rewards* (since some organizations do not offer *Rewards* for ISPC), *Threat Severity* (based on its relative low weighting of the PMT constructs), and *SETA* (as it likely to be correlated with *General Security Awareness* of the selected recent study).

3.3.5.3 Deterrence Theory

DT (Gibbs, 1975), rooted in the criminology discipline, helps us understand the effect of deterrent factors on illicit and/or other unwanted behaviours such as ISP noncompliance. DT proposes that the greater the *perceived severity, certainty,* and *celerity* (i.e., swiftness) of the punishment/sanction for the unwanted behaviour, the more the individual is deterred from performing the behaviour. This is another theory from which antecedents, selected as a result of applying *Steps 1–3* of our causal model procedure, were drawn. Next, we briefly discuss PMT.

While research demonstrates that individuals can be dissuaded from engaging in unwanted InfoSec-related behaviours when sanctions/punishment relevant to the unwanted behaviours are highlighted, the results of this research stream is mixed. For instance, some studies report no influence of sanctions on ISPCI (e.g., Moody et al., 2018; Rajab & Eydgahi, 2019), yet others found support for only some deterrent factors (e.g., Balozian, Leidner, & Warkentin, 2017; Li, Sarathy, Zhang, & Luo, 2014), and still others found support; however, the construct's influence may be negative or positive (e.g., D'Arcy, Hovav, & Galletta, 2009; Herath & Rao, 2009b). Despite these demonstrable inconsistencies, Cram et al. (2019) found that DT constructs have some predictive ability across a range of positive and negative IS security-related behaviours. Resulting from applying **Steps 1–3** of our strong, novel, and parsimonious causal model procedure, we incorporated *punishment severity* and *detection certainty* in our new causal model but not *punishment expectancy* and *resource vulnerability* (other deterrent factors); for parsimony sake as well as the fact that they explain limited amounts of variance in ISPC (see Table 3.1).

3.3.5.4 Information Security Awareness

Another relevant literature to be reviewed because of the antecedents selected from applying **Steps 1–3** of our causal model procedure is ISA. ISA refers to an individual's overall awareness (mindfulness), understanding, and knowledge of potential issues related to InfoSec and their ramifications (Bulgurcu et al., 2010; Donalds & Osei-Bryson, 2017). The InfoSec literature posits that ISA is a means of improving employees understanding and knowledge of potential InfoSec threats and risks they may likely face and appropriate actions to take to protect the organization's information assets (Donalds, 2015). It is reasonable to expect that increased ISA can persuade employees to go beyond prescriptions in the ISP, therefore improving ISPC.

Empirical results support ISA as a means for improving ISPC-related behaviours, such as reduction in IS misuse intention (D'Arcy et al., 2009), password compliance (Donalds & Osei-Bryson, 2020), actual cybersecurity compliance (Donalds, 2015; Donalds & Osei-Bryson, 2020), and ISPCI (Bulgurcu et al., 2010; X. Chen, Chen, & Wu, 2018). Results of this research stream affirm the beneficial effect of awareness on improving ISPC-related behaviours. Furthermore, Cram et al. (2019) confirm the explanatory power of awareness in predicting ISPC. Resultantly, from applying **Steps 1–3** in our causal model process, we included *general security awareness* in our new ISPC causal model.

3.3.5.5 General Security Orientation

An emergent factor cited as an important antecedent of ISPC is GSOR (see **Sub-Steps 2-a** and **2-b**). GSOR is mapped to an individual's 'security consciousness'; it refers to a user's predisposition and interest concerning practicing computer security (Ng, Kankanhalli, & Xu, 2009). Originating from the health belief model (HBM), GSOR is akin in perspective to the general health orientation variable of that model. General health orientation refers to the individual's predisposition or habit concerning health seeking behaviour, in general (Walker & Thomas, 1982). It attempts to capture an individual's generalized response tendency across a range of health preserving behaviours (Walker & Thomas, 1982). Since the HBM seeks to explain (preventative) health behaviour, we argue that it is applicable to the InfoSec context in that employees' appropriate or healthy security behaviours can be viewed as preventative behaviours to avert security incidents. Therefore, we assert that GSOR attempts to capture employees' generalized response tendency, or habit concerning healthy security behaviours, which could include reading security bulletins, exercising caution when clicking links/email attachments from unknown senders, using strong passwords, etc.

While GSOR has not received much attention in the context of InfoSec, there is burgeoning empirical evidence that this construct is important in explaining InfoSec-related behaviours. For instance, GSOR is found to have statistically significant influence on InfoSec compliance efficiency (Donalds, Osei-Bryson, & Samoilenko, 2019), employees' and users' general cybersecurity, and InfoSec compliance behaviour as well as employees' password security compliance behaviour (Donalds & Osei-Bryson, 2017, 2020) and has moderation effects with perceived severity on users' computer security behaviour (Ng et al., 2009). Since GSOR is not identified in Cram et al.'s (2019) MAR, but has been empirically validated in more recent studies than the MAR as a strong predictor of InfoSec compliance behaviour, we included *GSOR* in our new causal model (see **Steps 1–3**).

3.3.5.6 Personal Norms

Adapted from SBT(Hirschi, 1969), PN suggests that an individual's private and internalized norms, grounded in his/her beliefs and values, influence his/her deviant behaviour. PN is relevant to the study of ISPC because the decision to violate (or not) the ISP may be driven by the individual's innate feelings or desire, based on his/her own value systems and beliefs for compliance/non-compliance.

Extant research affirms that individuals' beliefs and values (i.e., their moral standards) impact their InfoSec-related behaviours or attitudes. Demonstrably, individuals' beliefs and values are found to influence their behaviour in obeying ISPs (Myyry, Siponen, Pahnila, Vartiainen, & Vance, 2009). PN has statistically significant influence on user commitment to volitional systems adoption and use (Malhotra & Galletta, 2005), employees' Internet use policy (Li et al., 2014) and attitude towards ISPC (Ifinedo, 2014; Safa, von Solms, & Furnell, 2016). *Personal norms* is included in our new causal model as it is identified in Cram et al.'s (2019) MAR as explaining the largest percentage of variance in InfoSec compliance (see Table 3.1), as per the guideline in our robust, novel, and parsimonious model generating procedure presented above. Consistent with the outputs from the various sub-steps in **Steps 1–3** above, Table 3.2 summarizes the main constructs used in our new robust, novel, parsimonious causal model, their associated theories/literatures, and definitions.

3.3.6 Logical Justification of the Hypotheses Resulting from Applying Steps 1–3

3.3.6.1 *Theory of Planned Behaviour*

The relationship between attitude and behavioural intention has been widely tested and supported in the IS literature. The TPB posits that attitude towards a behaviour is a strong predictor of an individual's intention to engage in that behaviour (Ajzen, 1991). Thus, we argue that employees with positive attitudes towards the ISP should have favourable tendencies to comply with its requirements. Furthermore, the ISP compliance literature provides evidence that attitude influences compliance behaviour (e.g., Bauer & Bernroider, 2017; Bulgurcu, Cavusoglu, & Benbasat, 2009; D'Arcy & Lowry, 2019; Ifinedo, 2012, 2014; Moody et al., 2018). Thus,

> **H1:** *Attitude* towards ISP compliance (positively) influences ISP *Compliance Intention.*

Recall, *Self-Efficacy* relates to an individual's belief in his/her capability to perform a task. The literature provides evidence of the relationship between *Self-Efficacy* and IS-related behaviours such as computer usage (Compeau & Higgins, 1995), exercising care with email attachments (Ng et al., 2009), intention to use anti-spyware application (Johnston & Warkentin, 2010), and general security compliance behaviour as well as password compliance behaviour (Donalds & Osei-Bryson, 2020). Further, it is expected that self-efficacy enhances individuals' effectiveness in executing IS-oriented tasks, thereby establishing more favourable attitudes towards use/compliance with them. For instance, Compeau, Higgins, and Huff (1999) found that self-efficacy has significant effects on individual's technological attitudes such as anxiety towards the technology and Herath and Rao (2009b) found that self-efficacy significantly influences security compliance attitude with ISP. Self-efficacy has also been found to have statistically significant influence on ISP compliance (e.g., Bulgurcu et al., 2010; D'Arcy & Lowry, 2019; Herath & Rao, 2009b; Ifinedo, 2012, 2014). Therefore, we argue that an individual who perceives himself/herself as having the competence to perform security-related actions that minimize the risks of security breaches/incidents is more likely to comply with requirements of the ISP and have more positive feelings towards the ISP. Thus,

H2a: Security *Self-Efficacy* (positively) influences ISP *Compliance Intention*.

H2b: Security *Self-Efficacy* (positively) influences *Attitude* towards ISP *Compliance*.

In this investigation, we posit that one's cognitive perception of his/her capability to perform or not the requisite behaviour may influence his/her perception that the given behaviour may or may not lead to the requisite outcome. This argument relates to *Self-Efficacy* and *Response Efficacy*. More specifically, we argue that if an employee does not understand the ISP well enough, s/he may not comprehend how complying with the ISP will help. On the other hand, when the employee has a better understanding of the ISP, then s/he may be better able to see how complying with the ISP may help protect the organisation IS assets. Given that prior research found statistically significant interaction effect for *Self-Efficacy* and *Response Efficacy* on intentions to adopt recommended preventative health behaviour, i.e., elimination or reduction of cigarette smoking (Maddux & Rogers, 1983), our proposed argument that self-efficacy has a relationship with response efficacy, seemingly is a reasonable one. Although no prior study has offered theoretical explanation for the link between self-efficacy and response efficacy, statistical evidence exists that suggest a link between these two variables. For instance, Wall, Palvia, and Lowry (2013) in a post hoc analysis test found that self-efficacy had a strong relationship with response efficacy. We therefore propose than an individual's cognitive perception of his/her ability to perform ISP security-related behaviours may improve his/her understanding of how complying with the ISP will help to reduce IS-related threats. Thus,

H2c: Security *Self-Efficacy* (positively) influences *Response Efficacy*.

The view that an individual's behaviour is likely influenced by the behavioural expectations of significant others (i.e., social influence) is consistent with findings in the IS literature. For instance, social influence was found to be an influential factor of an employee's intention to follow the USB-related

ISP guidelines (Aurigemma & Mattson, 2017). Workgroup norms was found to be a key determinant of end-user intentions to engage in non-malicious security violations in the workplace (Guo, Yuan, Archer, & Connelly, 2011). In considering the norms in an organisational setting, researchers have considered employees' perceptions of the expectations of managers, colleagues, and IS specialists with regard to ISP compliance (e.g., Bulgurcu et al., 2010; Herath & Rao, 2009b; Siponen et al., 2014). Consistent with these studies, we argue that this form of social influence, or normative belief, will have a persuasive effect on employees' ISP compliance intention. Thus,

H3: *Normative Belief* (positively) influences ISP *Compliance Intention*.

3.3.6.2 Protection Motivation Theory

Recall, the coping appraisal process consists of *Self-Efficacy, Response Efficacy,* and *Response Cost.* Since *Self-Efficacy* was discussed in the previous section, no further discussion ensues. In the context of IS security, *Response Efficacy* involves actions taken by an employee that they believe may reduce or neutralize security threats. For instance, to minimize security breaches, employees may decide not to write down passwords on a sticky note. In the ISP compliance context, *Response Efficacy* addresses employee's belief in whether the guidelines and procedures of the ISP, if adhered to, minimizes IS security incidents. We argue that when an individual is confident that a particular course of action will likely lead to positive outcomes, they feel more motivated to engage in that action and are more likely to do so. Prior investigations found that *Response Efficacy* positively influences ISP *Compliance Behaviour* (Ifinedo, 2012; Johnston & Warkentin, 2010; Johnston et al., 2015; Pahnila, Karjalainen, & Siponen, 2013; Wall et al., 2013). Thus,

H4: *Response Efficacy* (positively) influences ISP *Compliance Intention*.

Research suggests that individuals are often reluctant to adhere to recommended security practices, such as anti-malware software adoption, if they perceive that considerable opportunity costs (e.g., time, effort, trouble, money, embarrassment) will be expended towards the action (Lee & Larsen, 2009; Workman, Bommer, & Straub, 2008). According to Woon, Tan, and Low (2005), response cost has a negative relationship with behaviour as reducing the response cost will increase the likelihood of the respondent performing the recommended behaviour. IS research provide evidence that response cost negatively predicts security-related intentions, including ISP compliance intention (Boss, Galletta, Lowry, Moody, & Polak, 2015; Lee & Larsen, 2009; Vance et al., 2012). Consistent with prior research, we argue that if employees consider the cost of adhering to the requirements of the ISP to exceed perceived benefits, they are less likely to comply with it. Thus,

H5: *Response Cost* (negatively) influences ISP *Compliance Intention.*

3.3.6.3 Deterrence Theory

Research suggests that deterrence measures, such as the severity of punishment and detection certainty, are useful mechanisms to address IS security-related issues. For instance, Kankanhalli, Teo, Tan, and Wei (2003) reported that greater deterrent efforts positively influenced IS security effectiveness. Similarly, Straub (1990) found that visible and active deterrence efforts reduced IS security abuses. Like others, we also argue that deterrence measures can affect employees' ISP compliance intentions. For instance, if employees believe that their non-compliance with the ISP will be detected (detection certainty) and severe punishment (punishment severity) enforced, then these factors can influence the employees' behavioural intentions. That is, IS security may be enhanced when the potential for strong punishment, such as termination, criminal prosecution, and fines, as well as when the probability of detection of violations are high. Prior studies have empirically demonstrated the relationship between detection certainty and ISP compliance behaviour/intention (Herath & Rao, 2009a, 2009b; Li et al., 2014; Li et al., 2010; Siponen, Pahnila, & Mahmood, 2010) and sanction severity and ISP compliance behaviour/intention (Balozian et al., 2017; Cheng et al., 2013; Herath & Rao, 2009a, 2009b; Siponen et al., 2010). Thus,

H6: *Punishment Severity* (positively) influences ISP *Compliance Intention.*

H7: *Detection Certainty* (positively) influences ISP *Compliance Intention.*

3.3.6.4 Information Security Awareness

ISA has been highlighted as one of the most crucial components for achieving information security in organisations (Y. Chen, Ramamurthy, & Wen, 2015; D'Arcy et al., 2009; Siponen, 2000). According to Fishbein and Ajzen (1975), one way of producing change in human belief is through persuasive communication. We argue that through persuasive ISA mechanisms, such as IS security-related videos and newsletters highlighting security incidents, employees should become more cognizant of risks related to IS security, which should eventually translate into changed attitude. In fact, Dinev and Hu (2007) found that individuals' awareness of protective technologies against spyware played a central role in influencing their use of these technologies. Moreover, other studies have shown that ISA has a direct influence on employees attitudes towards IS security compliance (Bauer & Bernroider, 2017; Bulgurcu et al., 2010). Thus,

H8a: IS *Security Awareness* (positively) influences attitude towards ISP *Compliance Intention.*

In addition to attitude towards ISP compliance, we also postulate that IS *Security Awareness* influences security *Compliance Intention.* We argue that as IS *Security Awareness* improves employees IS security knowledge, so too should their security related behaviour. For instance, Stanton, Stam, Mastrangelo, and Jolton (2005) found that employees' increased password awareness was positively correlated with the quality of their password practices. Other recent studies found a direct link between awareness and actual security behaviour such as password compliance behaviour and general security compliance behaviour (Donalds & Osei-Bryson, 2020) and cybersecurity compliance behaviour (Donalds, 2015). In keeping with these findings, we posit that awareness alone could motivate an employee to act, regardless of whether s/he has formed a positive attitude. Thus,

H8b: IS *Security Awareness* (positively) influences ISP *Compliance Intention.*

We also posit that ISA influences security *Self-Efficacy.* Donalds (2015) suggests that ISA not only improve employees' understanding and knowledge of potential IS security risks but also the appropriate actions to take to minimize security risks. Likewise, Siponen (2000) suggests that awareness should minimize user-related faults/incidents. These suggestions relate to self-efficacy; they seemingly suggest that ISA influences employees' security self-efficacy, that is, the employee's ability to perform relevant and effective security-related actions due in part to awareness. While there is no study, to the best of our knowledge, that explores the link between ISA and self-efficacy, we concur with the suggestion that ISA influences security self-efficacy. We argue that since employees' have knowledge about potential issues related to IS security, employees' active engagement or participation in security-related activities such as security training, reading, and implementing security tips should influence their ability to perform security-related actions (i.e., self-efficacy). This reasoning is consistent with the findings of Stanton et al. (2005) who reported that awareness had positive impacts on users changing their passwords more frequently and using more difficult to guess passwords. Thus,

> H8c: IS *Security Awareness* (positively) influences security *Self-Efficacy*.

Practitioners continue to espouse the need for awareness campaigns to improve security-related behaviours. For instance, in its National Cybersecurity Strategy (Jamaica Houses of Parliament, 2015), awareness is cited as a critical factor for the successful realization of Jamaica's cybersecurity plan. Researchers too are cognizant of awareness's key role and has established its importance in reducing unwanted/illicit security behaviours such as computer/IS misuse intentions (Choi, Levy, & Anat, 2013; D'Arcy et al., 2009) and software piracy intentions (Peace, Galletta, & Thong, 2003). Peace et al. (2003) suggest that organisations increase employees' awareness of the potential severity of punishment and the certainty punishment to decrease the intentions of illicit behaviour such as software piracy. In a similar vein, we argue that raising security awareness of the fact that employees are caught and by providing information about the punishment for their violations, should deter ISP non-compliance. D'Arcy et al. (2009) provide empirical evidence for the positive influence of awareness on punishment severity and certainty of sanctions (i.e., detection certainty). Hence,

> H8d: IS *Security Awareness* (positively) influences ISP compliance *Punishment Severity*.
> H8e: IS *Security Awareness* (positively) influences ISP compliance *Detection Certainty*.

3.3.6.5 General Security Orientation

Recall that *GSOR* relates to an individual's predisposition, habit, or consciousness concerning healthy security behaviour. This description of *GSOR* suggests that an employee may decide to protect, or not, his/her organisation's information assets; however, this behaviour will largely depend on his/her own inclination towards the behaviour. We argue that employees who are more predisposed to practicing healthy security behaviours would

more likely respond in a more efficacious manner regarding the ISP requirements. On the other hand, employees who are less security conscious are
much less likely to conform to the requirements of the ISP. We further argue
that whether an employee will respond in a more, or less, efficacious manner regarding ISP requirements, is in part determined by his/her *GSOR*.
Empirical evidence indicates that employees' GSOR significantly and positively influences their general security compliance behaviour as well as their
password compliance behaviour (Donalds & Osei-Bryson, 2020). While no
prior study has examined the link between *GSOR* and security compliance
attitude (to our knowledge), we posit that *GSOR* influences security compliance *Attitude*. We argue that employees with higher levels of security
health consciousness will engage in seeking security health-related information from various sources (e.g., organisational, traditional, e-platforms)
to satisfy their security health-related information needs. It is a reasonable
argument that engaging in such security health-related information seeking activities should produce improvement in employees' belief, eventually leading to increased positive attitudes towards security compliance.
This reasoning is consistent with prior research that revealed that health
conscious individuals engage in more health information-seeking activities (Dutta-Bergman, 2004a, 2004b). Further, Ahadzadeh, Sharif, Ong, and
Khong (2015) found empirical support for the influence of health consciousness on health-related attitude towards Internet use for health information.
Similarly, we expect that an employee's GSOR (or health consciousness)
will positively influence his/her security compliance attitude. Based on the
preceding discussion, we propose the following:

H9a: *General Security Orientation (GSOR) (positively) influences ISP
Compliance Intention.*

H9b: *General Security Orientation (GSOR) (positively) influences
Attitude towards ISP Compliance Intention.*

3.3.6.6 Personal Norms

The view that an employee's ISP compliance, or noncompliance behaviour,
may be driven by their internalized norms, grounded in his/her beliefs and
values, is consistent with findings in the IS security literature (e.g., Myyry
et al., 2009; Yazdanmehr & Wang, 2016). This view implies that employees
follow their own belief system when making IS security-related decisions
(Myyry et al., 2009). PN has also been found to motivate system acceptance

and use in alignment with personal values and beliefs (Malhotra & Galletta, 2005). We argue that if an employee feels that following the ISP is the right thing to do, s/he will more likely comply with the ISP, that is, s/he will more likely have a positive ISP compliance intention. Prior research reveals that PN/values/beliefs is a statistically significant determinant of ISP compliance behaviour/intention (Cheng et al., 2013; Myyry et al., 2009; Yazdanmehr & Wang, 2016). We also propose that PN influences security compliance attitude. We argue that an employee with favourable security beliefs and values will likely form positive attitude towards ISP compliance. Prior research provides empirical support for the influence of PN on ISP compliance attitude (Ifinedo, 2014). Based on the preceding, we theorize that employees are motivated to comply with the ISP in alignment with their personal values and beliefs and that these values and beliefs will positively influence their attitude towards ISP compliance. Thus,

H10a: *Personal Norm* (positively) influences *Attitude* towards ISP *Compliance Intention*.

H10b: *Personal Norm* (positively) influences ISP *Compliance Intention*.

3.3.7 Step 4: Assess the Potential of the Current Candidate Model

Sub-Step 4-a: The scientist/research team should assess whether the evaluation of this m_{New3} has the potential to offer valuable contributions to the literature. An important consideration here is that m_{New3} should be novel (i.e., H_{New3} is not a subset of the hypotheses of any model that had been addressed in any single previous study).

Output: The model generated in Step 3 is novel as the process for generating it guaranteed not to involve a subset of the hypotheses of any model that had been addressed in any single previous study.

Sub-Step 4-b: If the research team determines that it has such potential, then m_{New3} should be added to M, the current set of novel candidate causal models that are to be considered for empirical evaluation.

3.4 CONCLUSION

This chapter demonstrates our novel process for generating relevant, robust, novel, and parsimonious causal models, applying it to the crucial domain of ISPC. To illustrate our approach, we build upon the work of Cram et al.

(2019), leveraging their MAR to identify top-ranked empirical predictors of ISPC. We augment these findings through additional literature review, highlighting the emergent antecedents of GSOR and general ISA.

By applying this process, we developed a comprehensive causal model for ISPC research, integrating both established and emergent predictors. Furthermore, we present brief reviews of relevant theoretical perspectives supporting these predictors, providing a robust logical foundation for the hypotheses within our ISPC model. This illustration showcases the effectiveness of our process in constructing meaningful and parsimonious models, offering valuable insights into the complex factors influencing ISPC behaviour and paving the way for future research.

REFERENCES

Ahadzadeh, A. S., Sharif, S. P., Ong, F. S., & Khong, K. W. (2015). Integrating health belief model and technology acceptance model: An investigation of health-related internet use. *Journal of Medical Internet Research, 17*(2), 1–17. https://doi.org/10.2196/jmir.3564

Ajzen, I. (1991). The theory of planned behavior. *Organizational Behavior and Human Decision Processes, 50*(2), 179–211.

Ajzen, I., & Fishbein, M. (1980). *Understanding attitudes and predicting social behaviour.* Englewood Cliffs, New Jersey: Prentice-Hall Inc.

Aurigemma, S., & Mattson, T. (2017). Deterrence and punishment experience impacts on ISP compliance attitudes. *Information & Computer Security, 25*(4), 421–436. https://doi.org/10.1108/ICS-11-2016-0089

Balozian, P., Leidner, D., & Warkentin, M. (2017). Managers' and employees' differing responses to security approaches. *Journal of Computer Information Systems, 59*(3), 197–210. https://doi.org/10.1080/08874417.2017.1318687

Bandura, A. (1977). Self-efficacy: Toward a unifying theory of behaviour change. *Psychological Review, 84*(2), 191–215.

Bauer, S., & Bernroider, E. W. N. (2017). From information security awareness to reasoned compliant action: Analyzing information security policy compliance in a large banking organization. *The Data Base for Advances in Information Systems, 48*(3), 44–68.

Becker, M. H. (1974). The health belief model and personal health behavior. *Health Education Monograph Series, 2*(4), 324–508.

Boss, S. R., Galletta, D. F., Lowry, P. B., Moody, G. D., & Polak, P. (2015). What do systems users have to fear? Using fear appeals to engender threats and fear that motivate protective security behaviors. *MIS Quarterly, 39*(4), 837–864.

Bulgurcu, B., Cavusoglu, H., & Benbasat, I. (2009, August 6–9). *Roles of Information Security Awareness and Perceived Fairness in Information Security Policy Compliance.* Paper presented at *the 15th Americas Conference on Information Systems,* San Francisco, California.

Bulgurcu, B., Cavusoglu, H., & Benbasat, I. (2010). Information security policy compliance: An empirical study of rationality-based beliefs and information security awareness. *MIS Quarterly, 34*(3), 523–548.

Chen, X., Chen, L., & Wu, D. (2018). Factors that influence employees' security policy compliance: An awareness-motivation-capability perspective. *Journal of Computer Information Systems, 58*(4), 312–324. https://doi.org/10.1080/08874417.2016.1258679

Chen, Y., Ramamurthy, K. R., & Wen, K.-W. (2015). Impacts of comprehensive information security programs on information security. *The Journal of Computer Information Systems, 55*(3), 11–19.

Cheng, L., Ying, L., Li, W., Holm, E., & Zhai, Q. (2013). Understanding the violation of IS security policy in organizations: An integrated model based on social control and deterrence theory. *Computers & Security, 39,* 447–459.

Choi, M. S., Levy, Y., & Anat, H. (2013). *The Role of User Computer Self-Efficacy, Cybersecurity Countermeasures Awareness, and Cybersecurity Skills Influence on Computer Misuse.* Paper presented at *the Eighth Pre-ICIS Workshop on Information Security and Privacy (WISP2013)*, Milan, Italy.

Compeau, D. R., & Higgins, C. A. (1995). Computer self-efficacy: Development of a measure and initial test. *MIS Quarterly, 19*(2), 189–211.

Compeau, D. R., Higgins, C. A., & Huff, S. (1999). Social cognitive theory and individual reactions to computing technology: A longitudinal study. *MIS Quarterly, 23*(2), 145–158. https://doi.org/10.2307/249749

Cram, W. A., D'Arcy, J., & Proudfoot, J. G. (2019). Seeing the forest and the trees: A meta-analysis of the antecedents to information security policy compliance. *MIS Quarterly, 43*(2), 525–554. https://doi.org/10.25300/MISQ/2019/15117

D'Arcy, J., & Lowry, P. B. (2019). Cognitive-affective drivers of employees' daily compliance with information security policies: A multilevel, longitudinal study. *Information Systems Journal, 29*(1), 43–69. https://doi.org/10.1111/isj.12173

D'Arcy, J., Hovav, A., & Galletta, D. (2009). User awareness of security countermeasures and its impact on information systems misuse: A deterrence approach. *Information Systems Research, 20*(1), 79–98.

Dinev, T., & Hu, Q. (2007). The centrality of awareness in the formation of user behavioral intention toward protective information technologies. *Journal of the Association for Information Systems, 8*(7), 386–408.

Donalds, C. (2015). *Cybersecurity Policy Compliance: An Empirical Study of Jamaican Government Agencies.* Paper presented at *the SIG GlobDev Pre-ECIS Workshop*, Munster, Germany.

Donalds, C., & Barclay, C. (2022). Beyond technical measures: A value-focused thinking appraisal of strategic drivers in improving information security policy compliance. *European Journal of Information Systems,* 1–16. https://doi.org/10.1080/09600 85X.2021.1978344

Donalds, C., & Osei-Bryson, K.-M. (2017). *Exploring the Impacts of Individual Styles on Security Compliance Behavior: A Preliminary Analysis.* Paper presented at *the SIG ICT in Global Development, 10th Annual Pre-ICIS Workshop*, Seoul, Korea.

Donalds, C., & Osei-Bryson, K.-M. (2020). Cybersecurity compliance behavior: Exploring the influences of individual decision style and other antecedents. *International Journal of Information Management, 51.* https://doi.org/10.1016/j.ijinfomgt.2019.102056

Donalds, C., Osei-Bryson, K. M., & Samoilenko, S. (2019, May 1–3). *Exploring the Impacts of Intrinsic Variables on Security Compliance Efficiency Using DEA and MARS.* Paper presented at *the IFIP International Federation for Information Processing 2019*, Dar es Salaam, Tanzania.

Dutta-Bergman, M. J. (2004a). Health attitudes, health cognitions, and health behaviors among internet health information seekers: Population-based survey. *Journal of Medical Internet Research, 6*(2), 1–7.

Dutta-Bergman, M. J. (2004b). Primary sources of health information: Comparisons in the domain of health attitudes, health cognitions, and health behaviors. *Health Communication, 16*(3), 273–288. http://doi.org/10.1207/S15327027HC1603_1

Fishbein, M., & Ajzen, I. (1975). *Belief, attitude, intention and behavior: An introduction to theory and research.* Reading, MA: Addison-Wesley.

Gibbs, J. P. (1975). *Crime, punishment, and deterrence.* New York: Elsevier

Guo, K. H., Yuan, Y., Archer, N., & Connelly, C. (2011). Understanding nonmalicious security violations in the workplace: A composite behavior model. *Journal of Management Information Systems, 28*(2), 203–236. https://doi.org/10.2753/MIS0742-1222280208

Herath, T., & Rao, H. R. (2009a). Encouraging information security behaviors: Role of penalties, pressures and perceived effectiveness. *DECISION Support Systems, 47*(2), 154–165.

Herath, T., & Rao, H. R. (2009b). Protection motivation and deterrence: A framework for security policy compliance in organisations. *European Journal of Information Systems, 18*(2), 106–125. https://doi.org/10.1057/ejis.2009.6

Hirschi, T. (1969). *Causes of delinquency.* Berkeley, CA: University of California Press.

Ifinedo, P. (2012). Understanding information systems security policy compliance: An integration of the theory of planned behavior and the protection motivation theory. *Computers & Security, 31*(1), 83–95. https://doi.org/10.1016/j.cose.2011.10.007

Ifinedo, P. (2014). Information systems security policy compliance: An empirical study of the effects of socialisation, influence, and cognition. *Information & Management, 51*(1), 69–79.

Jamaica Houses of Parliament. (2015). *The Cybercrimes Act, 2015 (Act No. 31 of 2015).* Retrieved from www.japarliament.gov.jm/attachments/article/341/The%20Cybercrimes%20Act,%202015-final%20No.31.pdf

Johnston, A. C., & Warkentin, M. (2010). Fear appeals and information security behaviors: An empirical study. *MIS Quarterly, 34*(3), 549–566.

Johnston, A. C., Warkentin, M., & Siponen, M. (2015). An enhanced fear appeal rhetorical framework: Leveraging threats to the human asset through sanctioning rhetoric. *MIS Quarterly, 39*(1), 113–134.

Kankanhalli, A., Teo, H.-H., Tan, B. C. Y., & Wei, K.-K. (2003). An integrative study of information systems security effectiveness. *International Journal of Information Management, 23*(2), 139–154.

Lee, Y., & Larsen, K. R. (2009). Threat or coping appraisal: Determinants of SMB executives' decision to adopt anti-malware software. *European Journal of Information Systems, 18*(2), 177–187. http://doi.org/10.1057/ejis.2009.11

Li, H., Sarathy, R., Zhang, J., & Luo, X. (2014). Exploring the effects of organizational justice, personal ethics and sanction on internet use policy compliance. *Information Systems Journal, 24*(6), 479–502. https://doi.org/10.1111/isj.12037

Li, H., Zhang, J., & Sarathy, R. (2010). Understanding compliance with internet use policy from the perspective of rational choice theory. *Decision Support Systems, 48*(4), 635–645. https://doi.org/10.1016/j.dss.2009.12.005

Lowry, P. B., & Moody, G. D. (2015). Proposing the control-reactance compliance model (CRCM) to explain opposing motivations to comply with organisational information security policies. *Information Systems Journal, 25*(5), 433–463. https://doi.org/10.1111/isj.12043

Maddux, J. E., & Rogers, R. W. (1983). Protection motivation and self-efficacy: A revised theory of fear appeals and attitude change *Journal of Experimental Social Psychology, 19*(5), 469–479.

Malhotra, Y., & Galletta, D. (2005). A multidimensional commitment model of volitional systems adoption and usage behavior. *Journal of Management Information Systems*, 22(1), 117–151.

Moody, G. D., Siponen, M. T., & Pahnila, S. (2018). Toward a unified model of information security policy compliance. *MIS Quarterly*, 42(1), 285–311.

Myyry, L., Siponen, M., Pahnila, S., Vartiainen, T., & Vance, A. (2009). What levels of moral reasoning and values explain adherence to information security rules? An empirical study. *European Journal of Information Systems*, 18(2), 126–139.

Ng, B.-Y., Kankanhalli, A., & Xu, Y. C. (2009). Studying users' computer security behavior: A health belief perspective. *Decision Support Systems*, 46(4), 815–825.

Pahnila, S., Karjalainen, M., & Siponen, M. T. (2013, June 18–22). *Information security behavior: Towards multistage models*. Paper presented at *the 17th Pacific Asia Conference on Information Systems*, Jeju Island, Korea.

Peace, A. G., Galletta, D. F., & Thong, J. Y. L. (2003). Software piracy in the workplace: A model and empirical test. *Journal of Management Information Systems*, 20(1), 153–177. https://doi.org/10.1080/07421222.2003.11045759

Ponemon Institute. (2023). *2023 Cost of Insider Threats Global Report*. Retrieved January 19, 2024 from URL https://ponemonsullivanreport.com/2023/10/cost-of-insider-risks-global-report-2023/

Rajab, M., & Eydgahi, A. (2019). Evaluating the explanatory power of theoretical frameworks on intention to comply with information security policies in higher education. *Computers & Security*, 80, 211–223. https://doi.org/10.1016/j.cose.2018.09.016

Rogers, R. W. (1975). A protection motivation theory of fear appeals and attitude change. *The Journal of Psychology*, 91(1), 93–114. https://doi.org/10.1080/00223980.1975.9915803

Rogers, R. W. (1983). Cognitive and physiological processes in fear appeals and attitude change: A revised theory of protected motivation. In J. T. Cacioppo & R. E. Petty (Eds.), *Social Psychophysiology: A Sourcebook* (pp. 153–176): The Guilford Press.

Safa, N. S., von Solms, R., & Furnell, S. (2016). Information security policy compliance model in organizations. *Computers & Security*, 56(C), 70–82.

Siponen, M. T. (2000). A conceptual foundation for organizational information security awareness. *Information Management & Computer Security Journal*, 8(1), 31–41.

Siponen, M. T., Mahmood, M. A., & Pahnila, S. (2014). Employees' adherence to information security policies: An exploratory field study. *Information & Management*, 51(2), 217–224. http://doi.org/10.1016/j.im.2013.08.006

Siponen, M. T., Pahnila, S., & Mahmood, M. A. (2010). Compliance with information security policies: An empirical investigation. *Computer*, 43(2), 64–71. https://doi.org/10.1109/MC.2010.35

Sommestad, T., Hallberg, J., Lundholm, K., & Bengtsson, J. (2014). Variables influencing information security policy compliance: A systematic review of quantitative studies. *Information Management & Computer Security*, 22(1), 42–75. https://doi.org/10.1108/IMCS-08-2012-0045

Sommestad, T., Karlzén, H., & Hallberg, J. (2015). The sufficiency of the theory of planned behavior for explaining information security policy compliance. *Information & Computer Security*, 23(2). https://doi.org/10.1108/ICS-04-2014-0025

Stanton, J. M., Stam, K. R., Mastrangelo, P., & Jolton, J. (2005). Analysis of end user security behaviors. *Computers & Security*, 24(2), 124–133.

Straub, D. W. (1990). Effective IS security: An empirical study. *Information Systems Research*, 1(3), 255–276.

Vance, A., Siponen, M. T., & Pahnila, S. (2012). Motivating IS security compliance: Insights from habit and protection motivation theory. *Information & Management*, 49(3–4), 190–198.

Walker, L. R., & Thomas, K. W. (1982). Beyond expectancy theory: An integrative motivational model from health care. *Academy of Management Review*, 7(2), 187–194. https://doi.org/10.5465/amr.1982.4285551

Wall, J. D., Palvia, P., & Lowry, P. B. (2013). Control-related motivations and information security policy compliance: The role of autonomy and efficacy. *Journal of Information Privacy and Security*, 9(4), 52–79. https://doi.org/10.1080/15536548.2013.10845690

Woon, I., Tan, G.-W., & Low, R. (2005). *A Protection Motivation Theory Approach to Home Wireless Security*. Paper presented at *the International Conference on Information Systems (ICIS)*, Las Vegas, NV.

Workman, M., Bommer, W. H., & Straub, D. (2008). Security lapses and the omission of information security measures: A threat control model and empirical test. *Computers in Human Behavior*, 24, 2799–2816. https://doi.org/10.1016/j.chb.2008.04.005

Yazdanmehr, A., & Wang, J. (2016). Employees' information security policy compliance: A norm activation perspective. *Decision Support Systems*, 92, 36–46.

4

Empirical Evaluation of the Causal Model of the Case Study

4.1 EVALUATION METHODOLOGY

To test the causal model presented in Chapter 3, Figure 3.1, we used a survey research design to collect data. In order to enhance validity and reliability, we adapted/adopted previously validated scales to measure the constructs in our model (Stone, 1978). We pre-tested our initial survey instrument, after which the final instrument was administered online.

4.2 INSTRUMENT DEVELOPMENT

The measurement scale for each construct was adapted/adopted from previously validated scales (see Table B.1 in Appendix B). We pretested the survey instrument using an expert panel of eight IT professionals and academics, who provided feedback to improve wording and validity. After making changes regarding wording, the panel experts agreed that the items were clearly written, realistic, and relevant, confirming face and content validity. The control variables were measured formatively; however, all other constructs were measured reflectively with multiple items, based mainly on a 5-point Likert-like scale, rated from 1 = "Strongly disagree" to 5 = "Strongly agree." We note that although a 5-point scale has been criticized by researchers, Dawes (2008, p. 1), in his study that addressed *"how using Likert-type scales with either 5-point, 7-point or 10-point format affects the resultant data in terms of mean scores, and measures of*

    DOI: 10.1201/9781032678931-4

dispersion and shape..." found that the 5- and 7-point scales produced the same mean score as each other, once they were rescaled to a comparable mean score out of 10... In terms of the other data characteristics, there was very little difference among the scale formats in terms of variation about the mean, skewness or kurtosis.

These results indicate "*good news*" for researchers; "*5- and 7-point scales can easily be rescaled with the resultant data being quite comparable*" (Dawes, 2008, p. 1). Also included in the instrument were items to capture respondents' demographics.

4.3 DATA COLLECTION

Data was collected from employees in a large financial institution in Jamaica. The survey was installed on the institution's Intranet and an email sent to all employees soliciting their participation. Of the 600 employees, 155 responses were received, yielding a response rate of 25.8%. Table 4.1 summarizes the socio-demographic information of the respondents.

TABLE 4.1

Socio-Demographic Characteristics of the Respondents

Category/ Subcategory	Count	Percent (%)	Category/ Subcategory	Count	Percent (%)
Sex:			*Organisation tenure:*		
Female	94	60.6	0–5 years	65	41.9
Male	61	39.4	6–10 years	19	12.3
			11–15 years	26	16.8
Age range:			16–20 years	18	11.6
20–29	46	29.7	21–30 years	15	9.7
30–39	39	25.2	31–40 years	12	7.7
40–49	35	22.6			
50–59	35	22.6	*Job level:*		
			Assistant Director	20	12.9
			Head of Department	3	1.9

(Continued)

TABLE 4.1 (Continued)

Category/ Subcategory	Count	Percent (%)	Category/ Subcategory	Count	Percent (%)
Education level:			Director	15	9.7
High school degree	11	7.1	Division Head and above	0	0.0
Undergraduate degree	58	37.4	Senior Director	2	1.3
Graduate degree	76	49.0	Line Staff/Entry Level	74	47.7
Other	10	6.5	Supervisor	35	22.6

4.4 STATISTICAL POWER ANALYSIS

An appropriate sample size is important to obtain sufficient statistical accuracy to detect effects of interest existing in the population. To determine the minimum required sample size N for our study, we conducted prospective estimations, that is, before data collection and analysis. We used two methods to determine statistical power: i) traditional Cohen's power tables for multiple regression (Cohen, 1992) and ii) a newer approach, the inverse square root method (Kock & Hadaya, 2018).

In the first prospective estimation, applying Cohen's power tables, sample size N is calculated as a function of the power level $1 - \beta$ (which indicates the probability of rejecting the false null hypothesis correctly); the significance level α (which is the researcher's long-term probability of erroneously rejecting the null hypotheses); the population effect size f^2 (which measures the effectiveness of a theory to explain or predict empirical observations [Webster & Starbuck, 1988] or reflects the magnitude of a phenomenon in a population [e.g., the impact of the independent variable on the dependent variable]); and the number of predictor variables of the equation containing the considered path coefficient/weight. By convention, effect sizes of 0.02, 0.15, and 0.35 are, respectively, termed small, medium, and large (Cohen, 1992). Statistical power $1 - \beta$ is usually set to 0.8, and the significance level α of 0.05 is assumed (Cohen, 1992). To determine N, we used G*Power 3.1 (Faul, Erdfelder, Lang, & Buchner, 2007), a high precision software tool to compute statistical power analyses, which implements Cohen's power guidelines. G*Power 3.1 indicated a minimum sample size of 127, assuming a medium effect size ($f^2 = 0.15$), power $1 - \beta$ of 0.8, significance level $\alpha = 0.05$, and number of predictors = 12.

Based on the inverse square root method, the minimum sample size N is estimated as the smallest positive integer that satisfies the following equation: $\widehat{N} > \left(\dfrac{2.486}{|\beta|_{min}} \right)$, where $|\beta|_{min}$ represents the minimum magnitude of the coefficient considered (Kock & Hadaya, 2018). In Excel, we used the following to calculate the minimum sample size: ROUNDUP ((2.486/0.215)^2,0). The inverse square root method suggests a minimum sample size of 134. Considering the outcomes of the two prospective estimations (i.e., Cohen's statistical power analysis and the inverse square root method), our sample size of 155 seems adequate to detect the effects of interests in our study.

4.5 COMMON METHOD BIAS

Common method bias (CMB) is caused by the method of collecting data rather than by the constructs and measurement items, which can lead to systematic error and wrong conclusion about the relationships between constructs (Podsakoff, MacKenzie, Podsakoff, & Lee, 2003). In order to mitigate against CMB, we observed Podsakoff et al.'s (2003) recommendations by applying several procedural remedies as follows: 1) we used different scale formats (5-point and 7-point Likert like scales); 2) we reduced the potential for social desirability as respondents remained anonymous; 3) we used different anchor points (strongly disagree to strongly agree and no extent to very great extent); and 4) improved scale items as an expert panel of eight IT professionals and academics provided feedback about wording of the questions as well as face and content validity. We also performed several post hoc statistical tests to examine the potential of CMB. First, a test of partial correlation was applied, in which a marker variable is introduced into the model to see if it varied the relationships in any way (Podsakoff et al., 2003) (see Table B.1 in Appendix B for the marker variable items). Our results show that no relationship was altered neither in their significance nor in their direction with and without the marker variable. Second, we inspected all latent variable correlations to ensure that they are below the 0.90 threshold (Pavlou, Liang, & Xue, 2007). As shown in Table 4.4, the highest correlation is 0.712. Third, we conducted Harman's single factor test. The results from this test showed that a single factor explained 32% of the total variance extracted, which is far less than the 50% cut-off value. Therefore, these statistical results, as well as the procedural remedies applied, indicate that CMB is not an issue in our investigation.

4.6 DATA ANALYSIS AND RESULTS

We used ADANCO 2.1 for Windows (http://www.composite-modeling. com/) (Henseler & Dijkstra, 2018) to perform partial least squares path modelling (PLS-PM) analysis of the research model. Using PLS-PM is advantageous as, according to Benitez, Henseler, Castillo, and Schuberth (2020), it has become a full-fledged variance-based estimator for structural equation modelling (SEM) that can deal with both reflective and causal-formative models. Further, enhancements to PLS-PM such as consistent PLS and the bootstrap-based test for overall model fit make PLS-PM suitable for causal research, that is, confirmatory and exploratory research (Benitez et al., 2020). Moreover, PLS-PM is an appropriate method for our study as researchers suggest that PLS-PM is appropriate when the structural model is complex and includes many constructs, indicators, and/or model relationships and when the path model includes one or more formatively measured constructs (Hair, Risher, Sarstedt, & Ringle, 2019); all of which are applicable to our investigation.

In ADANCO, as per Benitez et al. (2020), we used mode A to estimate the factor model, that is, reflective measurement models and the mode B weighting scheme for composite measurement models, that is, the control variables. Further, we used the factor weighting scheme for inner weighting and statistical inferences were based on the bootstrap procedure with 4,999 bootstrap runs (Benitez et al., 2020; Henseler, Hubona, & Ray, 2016).

4.7 ASSESSMENT OF REFLECTIVE AND COMPOSITE MEASUREMENT MODEL

4.7.1 Evaluation of Overall Fit of the Saturated Model

In order to assess the reflective and composite measurement models of our research model, we considered the guidelines as outlined in Benitez et al. (2020), Hair et al. (2019), and Henseler et al. (2016). First, we begun by evaluating the overall model fit with a saturated model, that is, with confirmatory factor/composite analysis. According to Benitez et al. (2020), the saturated model corresponds to a model in which all concepts are allowed to be freely correlated, whereas, the concept's operationalization is exactly as specified by the analyst. The overall model fit is useful to assess the validity of the measurement and composite models as potential model misfit can be attributed to misspecifications in the composite and/or measurement models (Benitez et al., 2020). Researchers have cautioned, however, that model fit measures as well as associated thresholds should be tentatively considered since a comprehensive assessment of these measures has not yet been conducted and that they should be examined in more detail in future methodological research (Benitez et al., 2020; Hair et al., 2019). Table 4.2 provides the overall model fit, that is, the discrepancy values and the 95% and 99% quantiles of their corresponding reference distribution.

While the value of the standardized root mean square residual (SRMR) is below the recommended threshold value of 0.08, there is contradictory results for the SRMR and one of the other tests of overall model fit, the unweighted least squares discrepancy (d_{ULS}), as shown in Table 4.2. Researchers recommend that all discrepancy values should be below the 95% quantile of the reference distribution (HI_{95}) (Benitez et al., 2020; Henseler et al., 2016). However, in the case of contradictory results, that is, when some or all of the discrepancy values are not below the 95% quantile (HI_{95}) of their corresponding distribution, scholars can evaluate whether the discrepancies are below the 99% quantile (HI_{99}) before rejecting the model (Benitez et al., 2020). As displayed in Table 4.2, the SRMR and the (d_{ULS}) discrepancy values are below the 99% quantile of their corresponding reference distributions (HI_{99}), while the geodesic discrepancy (d_G) value is below the 95% quantile (HI_{95}) of its corresponding distribution. Therefore, empirical evidence is obtained for all the latent variables in our research model.

TABLE 4.2

Results of the Confirmatory Factor Analysis

	Overall model fit evaluation			
Discrepancy	Value	HI_{95}	HI_{99}	Conclusion
SRMR	0.0660	0.0600	0.0661	Supported
d_{ULS}	4.1176	3.4042	4.1279	Supported
d_G	2.0299	2.1828	2.4074	Supported

4.8 ASSESSMENT OF THE REFLECTIVE MEASUREMENT MODEL

For reflective measurement models, item reliability, internal consistency reliability, convergent validity, and discriminant validity should be evaluated. Before discussing these measures, we note that normative belief is modelled as a construct comprising three reflective sub-constructs (see Figure 3.1). This approach deviates from previous approach that modelled normative belief as a formative (aggregate or reflective) latent construct (see Herath & Rao, 2009a; Herath & Rao, 2009b). We agree with these investigators that the indicators theoretically may be seen to employ different themes and may not be interchangeable; however, our view differs from theirs in how the construct is to be modelled. On inspecting the indicators, three themes emerged – management (BELF1 and BELF2), colleague (BELF3), and IT department/specialist (BELF4 and BELF5). Further, the indicators related to these sub-themes theoretically may be seen to measure its theme and are considered interchangeable. Thus, we model normative belief as three reflective sub-constructs and not as one formative construct.

The assessment of item reliability involves examining indicator loadings. Loadings above 0.708 are recommended, as they indicate that the construct explains more than 50% of the indicator's variance (Hair, Hult, Ringle, & Sarstedt, 2014). Except for two item loadings (COST1 and PNRM2), all other measurement item loadings on their respective constructs are above the minimum recommended threshold (see Table 4.3), thus providing acceptable item reliability. The two items with low loadings were subsequently removed from the model before further analysis.

Next, we assessed internal consistency reliability by examining Dijkstra–Henseler's ρ_A, Jöreskog's ρ_c, and Cronbach's α. A value of Dijkstra–Henseler's ρ_A larger than 0.707 is regarded as reasonable, as more than 50% of the variance in the construct scores can be explained by the latent variable (Benitez et al., 2020). Table 4.3 shows that the values for the Dijkstra–Henseler's ρ_A for the constructs in the research model range from 0.770 to 0.931, above the 0.707 threshold. Similarly, Jöreskog's ρ_c (Henseler et al., 2016) and Cronbach's α (Nunnally & Bernstein, 1994) values should have a minimum reliability of 0.7. As shown in Table 4.3, all Jöreskog's ρ_c and Cronbach's α values exceed the suggested 0.70 threshold.

TABLE 4.3

Measurement Model Results Summary

Construct	Code	Indicator Reliability	ρ_A	ρ_c	α	AVE
IS Security Compliance Intention	SINT1	0.603				
	SINT2	0.898	0.891	0.912	0.854	0.776
	SINT3	0.826				
IS Security Compliance Attitude	SATT1	0.932				
	SATT2	0.892	0.931	0.951	0.922	0.866
	SATT3	0.775				
IS Security Self-efficacy	SELF1	0.802				
	SELF2	0.752	0.884	0.917	0.866	0.787
	SELF3	0.800				
General IS Security Awareness	GSAW1	0.609				
	GSAW2	0.827	0.834	0.898	0.828	0.746
	GSAW3	0.802				
General IS Security Orientation	GSOR1	0.794				
	GSOR2	0.817	0.868	0.906	0.846	0.763
	GSOR3	0.677				
Response Cost	COST1	0.466[a]				
	COST2	0.854	0.874	0.915	0.819	0.844
	COST3	0.810				
Personal Norms	PNRM1	0.797				
	PNRM2	0.237[a]	0.786	0.901	0.781	0.820
	PNRM3	0.817				
Normative Belief	BELF1	0.807	0.770	0.897	0.770	0.813
	BELF2	0.818				
	BELF3	1.000	1.000	1.000		1.000
	BELF4	0.863	0.832	0.922	0.830	0.855
	BELF5	0.846				
Detection Certainty	DETC1	0.829				
	DETC2	0.891	0.916	0.920	0.871	0.794
	DETC3	0.661				
Punishment Severity	PUNS1	0.805				
	PUNS2	0.820	0.858	0.909	0.849	0.769
	PUNS3	0.683				
Response Efficacy	RESP1	0.737				
	RESP2	0.840	0.888	0.930	0.886	0.815
	RESP3	0.869				

Note:

[a] Indicates item was dropped.

These results indicate that all constructs in our research model have high reliability scores.

To evaluate convergent validity, two measures were assessed: the average variance extracted (AVE) (Chin, 1998; Fornell & Larcker, 1981) and indicator reliability (Hair et al., 2014). A minimum value of 0.5 is suggested for AVE, indicating that the corresponding latent variable explains at least 50% of the variance of its indicators (Fornell & Larcker, 1981). As seen in Table 4.3, all AVE values exceed 0.5. Indicator reliability represents how much of the variation in an item is explained by the construct and its suggested cut-off value is 0.5 (Hair et al., 2014). As shown in Table 4.3, all items, except the two items that were dropped, exceeded the minimum threshold. The results from these assessments provide a reasonable level of assurance of achieving convergent validity.

The final step in assessing the reflective measurement model is to evaluate discriminant validity. Discriminant validity is the extent to which a construct is empirically distinct from other constructs in the structural model. To obtain empirical evidence of discriminant validity, two traditional measures and one newer metric were considered, that is: i) the inspection of cross loadings and the examination of the Fornell–Larcker criterion (Fornell & Larcker, 1981) – two traditional measures; and ii) the examination of the heterotrait–monotrait ratio of correlations (HTMT), a more recent metric. As can be seen from the inspection of Table C.1 in Appendix C, each indicator loading is higher on its assigned construct than on all other constructs. According to the Fornell–Larcker criterion, the square root of the AVE of each construct should be higher than its correlation with other constructs. As can be seen in Table 4.4, the square root of each construct's AVE (along the diagonal) is higher than its intercorrelation with other constructs.

Recent research suggests, however, that inspecting cross-loadings and the Fornell–Larcker criterion are not sufficient for establishing discriminant validity. For instance, Henseler, Ringle, and Sarstedt (2015) demonstrate that the Fornell–Larcker criterion does not perform well, particularly when the indicator loadings on a construct differ only slightly. What Henseler et al. (2015) propose as a replacement for the Fornell–Larcker criterion and the cross-loading approach is HTMT. The HTMT is defined as the mean value of the item correlations across constructs relative to the (geometric) mean of the average correlations for the items measuring the same construct (Hair et al., 2019). According to

TABLE 4.4

Latent Variable Correlations

Construct	SINT	SATT	SELF	GSAW	COST	PNRM	RESP	PUNS	DETC	COLL	MGMT	IT-D	GSOR
Security Compliance Intention (SINT)	**0.881**												
Security Attitude (SATT)	0.712	**0.931**											
Security Self-Efficacy (SELF)	0.105	0.126	**0.887**										
General Security Awareness (GSAW)	0.389	0.409	0.241	**0.864**									
Response Cost (COST)	0.060	0.052	0.004	0.054	**0.919**								
Personal Norms (PNRM)	0.318	0.337	0.062	0.205	0.040	**0.906**							
Response Efficacy (RESP)	0.585	0.646	0.087	0.410	0.069	0.315	**0.903**						
Punishment Severity (PUNS)	0.107	0.107	0.059	0.082	0.001	0.178	0.097	**0.877**					
Detection Certainty (DETC)	0.134	0.192	0.043	0.079	0.001	0.159	0.196	0.304	**0.891**				
Normative Belief (Colleague – COLL)	0.241	0.251	0.140	0.210	0.050	0.172	0.285	0.163	0.163	–			
Normative Belief (Management – MGMT)	0.511	0.463	0.124	0.366	0.058	0.240	0.352	0.116	0.145	0.303	**0.902**		
Normative Belief (IT Dept. – IT-D)	0.326	0.342	0.122	0.238	0.020	0.188	0.270	0.176	0.156	0.462	0.512	**0.925**	
General Security Orientation (GSOR)	0.362	0.374	0.283	0.477	0.069	0.253	0.369	0.154	0.216	0.330	0.370	0.286	**0.873**

Note: The square root of the AVE is shown on the diagonal and is bolded. Normative belief – colleague is measured by a single indicator.

Benitez et al. (2020), the HTMT should be lower than 0.85 (a more strict threshold) or 0.90 (a more lenient threshold) or significantly smaller than 1. As seen in Table 4.5, the HTMT of all constructs are below the 0.90 threshold except security attitude with security intention to comply with ISP. While the threshold values can point to discriminant validity problems in research, according to Henseler et al. (2015, p. 128), "even if two constructs are highly, but not perfectly, correlated with values close to 1.0, the criterion is unlikely to indicate a lack of discriminant validity." As such, jointly, these tests suggest that discriminant validity has been established.

4.9 ASSESSMENT OF THE STRUCTURAL MODEL

Once the measurement model is deemed to be of sufficient quality, the next step is the assessment of the structural model. Structural model assessment includes the following: examination of the coefficient of determination (R^2), the path coefficient estimates, and their significance and effect sizes (f^2).

Figure 4.1 provides the results obtained from our PLS-PM analysis. To assess our structural model, similar to Johnston, Warkentin, and Siponen (2015) and (Siponen & Vance, 2010), we took a "staged" approach. In this staged approach, we first examined the influence of the control variables on employees' IS security compliance intention. In a subsequent stage of model testing, we added all the other variables as shown in our research model (see Figure 3.1).

Our control variables are gender, age, work experience, education, and job level. These variables were included to determine whether any of them have any effect on employees' IS security compliance intention. While the results from our PLS-PM analysis revealed that the control variables explained 9% of the variance in compliance intention, none of the variables had a statistically significant effect at this stage of the model testing.

In the next stage of the analysis, we added all the other variables to the previous model. As shown in Figure 4.1, the R^2 value of 0.784 indicates that the model explains a substantial amount of variance in *IS security compliance intention* (SINT). Further, the results show that 0.537 of the variance in employees' attitude towards IS security compliance, 0.241 of

TABLE 4.5

Heterotrait–Monotrait Ratio of Correlations (HTMT)

Construct	SINT	SATT	SELF	GSAW	COST	PNRM	RESP	PUNS	DETC	COLL	MGMT	IT-D
SATT	0.940											
SELF	0.379	0.386										
GSAW	0.735	0.729	0.580									
COST	0.282	0.257	0.070	0.284								
PNRM	0.676	0.680	0.293	0.561	0.251							
RESP	0.860	0.883	0.330	0.742	0.301	0.675						
PUNS	0.373	0.366	0.276	0.335	0.026	0.520	0.354					
DETC	0.411	0.478	0.228	0.321	0.018	0.474	0.489	0.647				
COLL	0.522	0.521	0.398	0.504	0.242	0.472	0.568	0.434	0.435			
MGMT	0.868	0.806	0.423	0.757	0.302	0.633	0.717	0.421	0.464	0.628		
IT-D	0.662	0.668	0.406	0.587	0.172	0.541	0.607	0.499	0.463	0.744	0.893	
GSOR	0.689	0.680	0.615	0.827	0.314	0.603	0.695	0.467	0.539	0.628	0.746	0.630

Note: SINT = Security Compliance Intention; SATT = Security Attitude; SELF = Security Self-Efficacy; GSAW = General Security Awareness; COST = Response Cost; PNRM = Personal Norms; RESP = Response Efficacy; PUNS = Punishment Severity; DETC = Detection Certainty; COLL = Normative Belief (Colleague); MGMT = Normative Belief (Management); IT-D = Normative Belief (IT Dept.); GSOR = General Security Orientation.

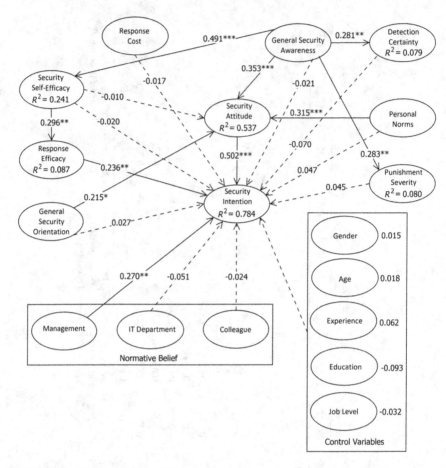

FIGURE 4.1
PLS-PM Results of the Structural Model.

Note: **Dashed lines represent unsupported paths; * $p < 0.05$; ** $p < 0.01$; *** $p < 0.001$.**

the variance in *security self-efficacy* (SELF), 0.087 of the variance for *response efficacy* (RESP), 0.079 of the variance in *detection certainty* (DETC), and 0.080 of the variance in *punishment severity* (PUNS) are explained by the variables considered in the model.

The results also provide evidence of the fully mediating role of *IS security attitude* (SATT) between *general security orientation* (GSOR), *general security awareness* (GSAW), and PNRM with SINT. A variable functions as a dominant mediator when Path *c* (independent variable [IV] to dependent variable) is reduced to 0, while Paths *a* and *b* (IV to mediator variable [Path *a*] and mediator variable to dependent variable [Path *b*]) are statistically significant (Baron & Kenny, 1986). As shown in Figure 4.1, the links between GSOR and SINT; GSAW and SINT; and PNRM and SINT are insignificant (Path *c* reduced to 0), providing no support for H9a, H8b, and H10b, respectively. However, Path *a* (GSOR to SATT, GSAW to SATT, and PNRM to SATT) is statistically significant, providing empirical support for H9b, H8a, and H10a, respectively. Similarly, Path *b*, SATT to SINT, is statistically significant; thus, H1is supported. These results confirm that SATT is a strong predictor of SINT.

The results also confirm that GSAW plays a vital role in organisational information security. As shown in Figure 4.1, GSAW to *security self-efficacy* (SELF), GSAW to *punishment severity* (PUNS), and GSAW to *detection certainty* (DETC) are statistically significant, providing support for H8c, H8d, and H8e, respectively.

The results also suggest that H2c (*response efficacy* [RESP] to SELF) is supported. The statistically significant paths between RESP and SINT and *normative belief-management* (MGMT) and SINT provide empirical support for H4 and H3a, respectively. These results suggest that employees believe that adhering to IS security rules/guidelines is an effective mechanism to prevent a threat to the organisational IS assets; and pressure from management positively influence their ISP compliance intention. The support for H2c when combined with the support for H4 (RESP to SINT) indicates that RESP mediates the relationship between SELF and SINT.

The results do not provide adequate statistical evidence that SELF influences SATT or SINT directly; thus, H2a and H2b are unsupported. Likewise for the following hypotheses: H3b –*normative belief* – *IT department* (IT-D) to SINT; H3c – *normative belief* – *colleague* (COLL) to SINT; H5 – *response cost* (COST) to SINT; H6 – PUNS to SINT; and H7 – DETC to SINT. Table 4.6 provides a summary of the results.

According to recent research (e.g., Benitez et al., 2020; Henseler et al., 2016), the effect sizes of the relationships between the constructs should be evaluated. The effect size is a measure of the magnitude of an effect that is independent of sample size. The f^2 values can be used to indicate small ($0.02 \leq f^2 < 0.15$), medium ($0.15 \leq f^2 < 0.35$), and large ($f^2 \geq 0.35$) effects, respectively (Cohen, 1992). According to Benitez et al. (2020, p. 11), "*just as all actors in a movie cannot play a leading role, it is unusual and unlikely that most constructs will have a large effect size in the model*". As shown in Table 4.6, the f^2 values in our sample for the hypothesized relationships range from 0.043 to 0.318 (small to medium) for the statistically significant relationships.

TABLE 4.6

Summary Results of the Structural Model Assessment

Hypothesis		β	*t*-value	Supported	f^2
H1:	**IS security attitude → ISP compliance intention**	0.502	5.381	**Yes**	0.301
H2a:	IS security self-efficacy → ISP compliance intention	−0.020	−0.469	No	0.001
H2b:	IS security self-efficacy → IS security attitude	−0.010	−0.176	No	0.000
H2c:	**IS security self-efficacy → Response efficacy**	0.296	3.260	**Yes**	0.096
H3a	**Norm. belief (Management) → ISP compliance intention**	0.270	2.744	**Yes**	0.108
H3b:	Norm. belief (IT-Dept.) → ISP compliance intention	−0.052	−0.632	No	0.004
H3c:	Norm. belief (Colleague) → ISP compliance intention	−0.024	−0.379	No	0.001
H4:	**Response efficacy → ISP compliance intention**	0.236	2.580	**Yes**	0.072
H5:	Response cost → ISP compliance intention	−0.017	−0.389	No	0.001
H6:	Punishment severity → ISP compliance intention	0.045	0.873	No	0.006
H7:	Detection certainty → ISP compliance intention	−0.070	−1.202	No	0.012
H8a:	**General security awareness → IS security attitude**	0.353	4.166	**Yes**	0.131
H8b:	General security awareness → ISP compliance intention	−0.020	−0.228	No	0.001
H8c:	**General security awareness → IS security self-efficacy**	0.491	5.836	**Yes**	0.318
H8d:	**General security awareness → Punishment severity**	0.283	3.267	**Yes**	0.087
H8e:	**General security awareness → Detection certainty**	0.281	2.757	**Yes**	0.086
H9a:	General security orientation → ISP compliance intention	0.027	0.319	No	0.001
H9b:	**General security orientation → IS security attitude**	0.215	2.502	**Yes**	0.043
H10a:	**Personal norms → IS security attitude**	0.315	3.789	**Yes**	0.155
H10b:	Personal norms → ISP compliance intention	0.047	0.843	No	0.006

Note: Norm = Normative.

4.10 DISCUSSION OF FINDINGS

Research on end user IS security compliance have offered many well-reasoned and empirically validated models with different sets of direct and indirect statistically significant determinants (antecedents) of ISP compliance intention. One approach to such a situation is to view these resulting models as competitors; another approach is to view them as offering descriptions of legitimate alternate path to achieving end user IS security compliance, some of which are most relevant for a given context.

In Chapter 3, we proposed a new causal model that included some constructs and hypotheses that were previously proposed in ISP compliance research as well as some new ones. As is typical in many IS behavioural science research (BSR) projects, empirical analysis of our proposed model resulted in some of our hypotheses achieving adequate statistical support while others did not, with the latter including some that had achieved statistical support in at least one other previous study, though in some cases there were mixed results (see Appendix D). It should be noted that whether a given hypothesis is supported in a given study could be determined by the presence of other variables and hypothesized causal links (e.g., Balozian, Leidner, & Warkentin, 2017; Siponen & Vance, 2010) in the current model.

Below, we provide discussion on some possible reasons for differences between the results of this study and that of previous studies. It should be noted that in doing this, we take the position that there could be multiple valid explanations for the same phenomenon.

4.11 DIRECT ANTECEDENTS OF IS SECURITY ATTITUDE?

Hypotheses H8a, H9b, and H10a each achieved sufficient statistical support in this study. It should be noted that while H8a and H10a were proposed and empirically supported in previous studies, the current study is the first that has explored H9a, which also obtained empirical support. While hypothesis H2b did not achieve adequate statistical support in this study, it had achieved statistical support in at least one previous study (see Table 4.7).

TABLE 4.7

Antecedents of IS security *Attitu de*: Case Study vs. Prior Research

	Any Previous Study		This Research
Hypothesis	**Supported**	**Rejected**	**Supported**
H2b: IS security self-efficacy → IS security attitude	✓	✓	No
H8a: General security awareness → IS security attitude	✓		Yes
H9b: General security orientation → IS security attitude			Yes
H10a: Personal norms → IS security attitude	✓		Yes

One possible explanation for the lack of observed effects of *IS self-efficacy* on *IS security attitude* pertains to our research setting. We surveyed employees employed by a public sector organisation engaged in highly procedural and bureaucratic work. The fact that the study's participants self-reported as having a moderate–high level of competence in performing security related actions (i.e., average *self-efficacy* is 3.7, on a 5-point scale) might also be one reason why there were no observed effects. This argument is consistent with and parallels a finding of Dinev, Jahyun, Qing, and Kichan (2009), who reported that perceived ease of use of the technology had a significant role in forming a positive attitude towards it for their South Korean respondents but not for their US respondents, who had higher self-reported computer skills (i.e., self-efficacy). Further, it is likely that the highly procedural and bureaucratic nature of the work nullified the effect of the employees' *IS self-efficacy* on their *IS security attitude*.

4.12 DIRECT ANTECEDENTS OF ISP COMPLIANCE INTENTION?

As shown in Table 4.8, hypothesis H1 achieved adequate statistical support. However, some of the relevant hypotheses which did not achieve adequate statistical support in this study have had mixed results in previous studies (i.e., H2a, H4, H5, H6, H7, and H9a), while others (i.e., H2a, H8b, and H10b) had achieved adequate statistical support in previous studies.

TABLE 4.8

Antecedents of ISP Compliance Intention: Case Study vs. Prior Research

	Any Previous Study		This Research
Hypothesis	Supported	Rejected	Supported
H1: IS security attitude → ISP compliance intention	✓	✓	Yes
H2a: IS security self-efficacy → ISP compliance intention	✓	✓	No
H3a Norm. Belief (Management) → ISP compliance intention	✓		Yes
H3b: Norm. Belief (IT-Dept.) → ISP compliance intention			No
H3c: Norm. Belief (Colleague) → ISP compliance intention			No
H4: Response efficacy → ISP compliance intention	✓	✓	Yes
H5: Response cost → ISP compliance intention	✓	✓	No
H6: Punishment severity → ISP compliance intention	✓	✓	No
H7: Detection certainty → ISP compliance intention	✓	✓	No
H8b: General security awareness → ISP compliance intention	✓		No
H9a: General security orientation → ISP compliance intention	✓	✓	No
H10b: Personal norms → ISP compliance intention	✓		No

We next discuss results that had mixed statistically significant results in previous studies but were not supported in this study.

4.12.1 Previous Studies – Mixed Results; This Study – Results Not Supported

As shown in Table 4.8, H2a, H5–H7, and H9a had mixed results (i.e., the hypotheses were statistically significant as well as statistically insignificant) in previous studies but were statistically insignificant in this study.

- H2a – In this study, the lack of statistical significance of *security self-efficacy* on *ISP compliance intention* can be because informed employees (i.e., about the importance of IS security) may comply with the ISP not because s/he perceives that s/he has the ability to perform requisite security actions that may minimize security threats but because s/he perceives that there are real security threats to the organisation IS assets. In line with this argument, Dinev and Hu (2007) indicated that an individual may feel compelled to use anti-virus and/or anti-spyware technologies regardless of the confidence s/he has in using them. Another plausible explanation for the difference in results could be attributed to the sample demographics. Our sample consisted of highly educated government employees engaged in highly procedural and bureaucratic work. Thus, the highly procedural and bureaucratic nature of the work may reduce the importance of *self-efficacy*. Further, in other studies where *self-efficacy* was statistically significant (e.g., Herath & Rao, 2009b; Johnston & Warkentin, 2010), the participants were more diverse, consisting of students, faculty, and university and business employees. Moreover, the demographic factors of the above cited studies varied more than in this study. This suggests that the difference in the result could be due to the WHERE dimension (e.g., organizational culture of context). According to Wall, Palvia, and Lowry (2013, p. 70), "*it may be that for certain populations, self-efficacy does not have a direct effect on compliance intentions*", supporting our suggestion that the WHERE dimension may account for *self-efficacy's* lack of statistical significant direct impact on *ISP compliance intention* in this study.

- H5 – One possible reason why the *response cost* to *ISP compliance intention* link is not supported in this study may be employees' low level of perceived inconvenience of adhering to the ISP. In this study, the average *response cost* is 2.4, on a 5-point scale. In other words, employees do not consider the inconvenience of adhering to IS security policies a legitimate reason for not complying with the policies.
- H6 – In this study, the lack of statistical significance of perceived *punishment severity* concurs with the findings of many studies in criminology and IS security compliance (see Balozian et al., 2017; Li, Sarathy, Zhang, & Luo, 2014; Paternoster & Simpson, 1996; Wenzel, 2004). There are several reasons why this hypothesis may not be supported. Similar to the argument presented for H5, one possible reason may be the somewhat low level of perceived punishment severity for ISP non-adherence. That is, the consequences of punishment relating to ISP non-adherence are not perceived by employees to be severe. This argument is in line with the results of a

study by Li et al. (2014); the result is that employees had a low level of perceived sanction [punishment] severity for Internet abuses. Additionally, it is plausible that *punishment severity* is likely to have less effect when influences of IS security is driven by expectations of others, such as management support for IS security, as was indicated by Cuganesan, Steele, and Hart (2018) and was found to be the case in this study (*normative belief – management*). Another reason could be that of "competition" from other constructs. In a post hoc analysis, we explored if removing *IS security attitude* from the model would result in the *punishment severity* to *ISP compliance intention* link being statistically significant; the result is that this link was then found to be statistically significant. It should be noted that this situation where the addition of a particular construct to a model results in the newly included construct being a statistically significant antecedent of a given target variable, but with some of the other previously significant antecedents no longer achieving statistical significance, has been previously reported in the IS security literature (e.g., Balozian et al., 2017; Siponen & Vance, 2010). It should, however, also be noted that the model that included *IS security attitude* had a higher R^2 value, providing better explanatory power. The result of the post hoc analysis also suggests that the difference in results could be explained by the WHATs (e.g., presence or absence of the *IS security attitude* construct) and HOWs (sets of hypotheses).

- H7 – Unlike our study, the studies in which this hypothesis achieved adequate statistical support either involved participants from multiple types of organisations or did not include the *IS security attitude* construct. Further, Balozian et al. (2017, p. 5) observed that "*the severity and certainty of sanctions tested alone are significant; however, when the whole model is tested together, the coercive approach loses its significance*". Similar to H6, some possible reasons for the difference in the results include differences in the causal models (i.e., the WHATs [e.g., presence or absence of the IS *security attitude* construct] and HOWs [sets of hypotheses]) and the context in terms of the WHERE (e.g., national culture and organisational culture).
- H9a – Both this study and that of Donalds and Osei-Bryson (2020) involved study participants who were located in the same country. However, while the sample of Donalds and Osei-Bryson's (2020) study involved individuals who were employed in a variety of industries, the participants of this study were employed by a bureaucratic

public sector organisation. This suggests that the difference in the result could be due to the WHERE dimension of context, but not in the sense of national culture but of organisational culture.

4.12.2 Previous Studies – Results Supported; This Study – Results Not Supported

In this section, we discuss results (H8b and H10b) which had statistically significant results in previous studies but were not supported in this study (Table 4.8).

- H8b – In the Donalds and Osei-Bryson (2020) study, the *general security awareness* to *ISP compliance intention* link was significant. While the participants of this and that of Donalds and Osei-Bryson (2020) studies were located in the same country, the Donalds and Osei-Bryson (2020) causal model did not include *IS security attitude* as a mediator variable. Again, we conducted a post hoc analysis to explore whether removing *IS security attitude* from the model would result in the *general security awareness* to *ISP compliance intention* link being statistically significant. The link between *general security awareness* and *ISP compliance intention* was then found to be statistically significant. Thus, the difference in result could be explained by the WHATs and HOWs since *IS security attitude* was found to be a fully mediating variable for *general security awareness* and other variables in this study. Moreover, given the fact that *IS security attitude* is a fully mediating variable, our results could be considered to empirically indicate that *general security awareness* indirectly impacts *ISP compliance intention*.
- H10b – There are several reasons why the link between *personal norms* and *ISP compliance intention* was not supported. One possible explanation is that employees may not comply with the rules outlined in the ISP because they do not perceive it in their best interest to do so. For instance, employees may decide to violate the information security policy that one should not share their passwords with co-workers since it would not uphold cooperation between friends. A similar finding was reported by Myyry, Siponen, Pahnila, Vartiainen, and Vance (2009). Another possible reason why *personal norms* is not statistically significant in this study may be because ISP compliance is mandatory and not volitional. According to Malhotra

and Galletta (2005), personal norms tend to motivate system adoption and volitional system use in alignment with personal values and beliefs. Drawing on this finding, we posit that since ISP compliance is not volitional but mandatory, employees' personal values and beliefs do not influence their compliance intention; that is, employees are mandated to comply with security rules/guidelines. In studies where the link from *personal norms* to *ISP compliance intention* was statistically significant (see Li et al., 2014; Li, Zhang, & Sarathy, 2010; Yazdanmehr & Wang, 2016), *IS security attitude* was not included in the causal model. Similarly, in studies where the link from *personal norm* to *IS security attitude* was statistically significant, the link between *personal norm* and *ISP compliance intention* was not investigated (see Ifinedo, 2014; Safa, von Solms, & Furnell, 2016). In this study, we investigated the links from *personal norms* to *IS security attitude* and *ISP compliance intention*. Therefore, in another post hoc analysis test with *IS security attitude* removed from the causal model, the result is that the link from *personal norms* to *ISP compliance intention* became statistically significant. Thus, the difference in results could also be explained by the *WHATs* and the *HOWs*.

4.13 DIRECT ANTECEDENTS OF THE OTHER EXTRINSIC VARIABLES?

Table 4.9 shows that hypotheses H8c–H8e each involves *general security awareness* as the antecedent, and each achieved sufficient statistical support in this study. It should be noted that while H8d and H8e were proposed and empirically supported in a previous study, the current study is the first that has explored H8c, which also obtained empirical support. It should also be noted that empirical support was found for H2c. This is important since as mentioned prior, although the direct *IS security self-efficacy* to *ISP compliance intention* link did not receive statistical support in this study, and since the *IS security self-efficacy* to *response efficacy* and *IS response efficacy* to *ISP compliance intention* links both achieved statistical support in this study, our results support the position that *IS security self-efficacy* has at least an indirect impact on *ISP compliance intention*.

TABLE 4.9

Summary Results of Other Relationships in This Study and Prior Research

		Any Previous Study		This Research
Hypothesis		**Supported**	**Rejected**	**Supported**
H2c:	Response efficacy → IS security self-efficacy	✓[a]		Yes
H8c:	General security awareness → IS security self-efficacy			Yes
H8d:	General security awareness → Punishment severity	✓		Yes
H8e:	General security awareness → Detection certainty	✓		Yes

[a] No previous study provided theoretical explanation for the link between response efficacy and IS security self-efficacy; however, a post hoc analysis test found support for the relationship.

4.14 CONTRIBUTION TO THEORY

Gregor (2006) suggests that there are several theory types, including explanatory theory and predictive theory. An explanatory theory that is in the format of a causal model (e.g., Whetten, 1989) consists of the WHATs (concepts), HOWs (links between the concepts), WHYs (justification for each link), and the WHERE and WHEN (i.e., the situations in which the theory applies). On the other hand, a predictive theory indicates *"what will happen in the future if the condition hold"* Gregor (2006, p. 619). Typically, quantitative empirical IS research involves the explanatory type, but often without making plain the context (i.e., WHERE and WHEN) in which the theory is relevant. The theoretical contributions of this study involve these two types: explanatory and predictive.

4.15 CONTRIBUTION TO EXPLANATORY THEORY

This study involves the development and empirical analysis of a proposed causal model. Some contributions of our empirical analysis to explanatory theory are as follows:

- *ISP compliance intention*: This study identifies statistically significant *ISP compliance intention* antecedents as *IS security attitude, response efficacy*, and *normative belief (management)*, with the first being a fully mediating variable and the others independent extrinsic variables. This study demonstrates that with respect to *ISP compliance intention*, high explanatory power can be achieved by a model that does not involve as its statistically significant antecedents variables such as *punishment severity, detection certainty, response cost*, and *IS security self-efficacy* and some of the other variables identified in previous studies as statistically significant antecedents.

- *IS security attitude*: The identified statistically significant predictors of *IS security attitude* are *general security awareness, general security orientation*, and *personal norms*. This study demonstrates that with respect to *IS security attitude*, high explanatory power can be achieved by a model that does not involve as its statistically significant antecedents variables such as *IS security self-efficacy* and some of the other variables identified in previous studies as statistically significant antecedents.

4.16 CONTRIBUTION TO PREDICTIVE THEORY

We executed another PLS-PM analysis with an adjusted structural model that only included the links that were found to be statistically significant in our original structural model. Given that even links that are not statistically significant contributed to the R^2 values, our aim was to determine whether the adjusted structural model still had high R^2 values with respect to the statistically significant antecedents of *ISP compliance intention* and *IS security attitude*. The results (see Figure 4.2) include one of the highest reported R^2 for *ISP compliance intention* (i.e., 0.764) and one of the highest for *IS security attitude* (i.e., 0.535) in the literature. This adjusted structural model (WHATs and HOWs) could thus reasonably be considered a strong model for predicting *ISP compliance intention*. Given that justificatory arguments (i.e., the WHYs) were provided for each of the relevant statistically significant links, then the results shown in Figure 4.2 can reasonably be considered to offer a predictive theory.

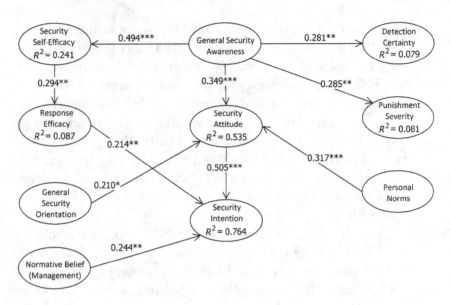

FIGURE 4.2
PLS-PM Predictive Theory Model.

4.17 CONCLUSION

The aim of this chapter was to evaluate the causal model presented in Chapter 3, which was generated from applying our relevant causal model generating process (see Chapter 2). To test the hypotheses in the causal model (Figure 3.1), we collected data via a survey instrument that was administered online. The measurement items for each construct were either adapted or adopted from previously validated scales. The survey instrument was pre-tested by IT professionals and academics. While each latent variable was measured reflectively, all control variables were measured formatively. The results from our sample size tests supported our sample size of 155, which indicated that the same was adequate to detect the effects of interests in our investigation. Our tests of CMB also indicated that CMB was not an issue in our investigation.

Results from our PLS-PM analysis revealed that the model had good overall model fit, supporting the latent variables presented in our causal model. Tests of reliability, internal consistency reliability, convergent validity, and discriminant validity all exceeded their recommended thresholds, indicating that the measurement model of our causal model was supported.

Results from the assessment of our structural model showed strong support for our causal model. Overall, the model explained a substantial amount of variance in the dependent variable, *ISP compliance intention* (SINT), with an R^2 value of 0.784. The results also provided evidence of the fully mediating role of *IS security attitude* (SATT) between *general security orientation* (GSOR), *general security awareness* (GSAW), and *personal norms* (PNRM) with SINT. However, the results did not provide adequate statistical support for the direct influence of *self-efficacy* (SELF), *response cost* (COST), *punishment severity* (PUNS), and *detection certainty* (DETC) on SINT.

In this investigation, we proposed a new causal model that included some latent variables and hypotheses that were previously proposed in ISP compliance research as well as some new ones. As is typical in many BSR projects, empirical analysis of our proposed model resulted in some of our hypotheses achieving adequate statistical support while others did not, with the latter including some that had achieved statistical support in at least one other previous study, though in some cases there were mixed results. Notwithstanding, the results shown in Figure 4.1 reveal one of the highest reported R^2 values for *ISP compliance intention* (i.e., 0.784), and one of the highest for *IS security attitude* (i.e., 0.537) in the literature. Thus, based on the empirical results, our proposed causal model can reasonably be considered a strong model for ISP *compliance intention*; which demonstrated and validated our proposed causal model generating process and our claims that when applied, can generate relevant causal models that are not only robust, novel, and parsimonious, but each of which is likely to offer strong empirical support with respect to the given dependent variable.

REFERENCES

Balozian, P., Leidner, D., & Warkentin, M. (2017). Managers' and employees' differing responses to security approaches. *Journal of Computer Information Systems, 59*(3), 197–210. https://doi.org/10.1080/08874417.2017.1318687

Baron, R. M., & Kenny, D. A. (1986). The moderator-mediator variable distinction in social psychological research: Conceptual, strategic, and statistical considerations. *Journal of Personality and Social Psychology, 51*(6), 1173–1182.

Benitez, J., Henseler, J., Castillo, A., & Schuberth, F. (2020). How to perform and report an impactful analysis using partial least squares: Guidelines for confirmatory and explanatory IS research. *Information & Management, 57*(2), 103168. https://doi.org/10.1016/j.im.2019.05.003

Chin, W. W. (1998). Issues and opinion on structural equation modeling. *MISQ*, *22*(1), vii–xvi.

Cohen, J. (1992). Quantitative methods in psychology: A power primer. *Psychological Bulletin*, *112*(1), 155–159.

Cuganesan, S., Steele, C., & Hart, A. (2018). How senior management and workplace norms influence information security attitudes and self-efficacy. *Behaviour & Information Technology*, *37*(1), 50–65. https://doi.org/10.1080/0144929X.2017.1397193

Dawes, J. (2008). Do data characteristics change according to the number of scale points used? An experiment using 5-point, 7-point and 10-point scales. *International Journal of Market Research*, *50*(1), 61–104.

Dinev, T., & Hu, Q. (2007). The centrality of awareness in the formation of user behavioral intention toward protective information technologies. *Journal of the Association for Information Systems*, *8*(7), 386–408.

Dinev, T., Jahyun, G., Qing, H., & Kichan, N. (2009). User behaviour towards protective information technologies: The role of national cultural differences. *Inormation Systems Journal*, *19*(4), 391–412. https://doi.org/10.1111/j.1365-2575.2007.00289.x

Donalds, C., & Osei-Bryson, K.-M. (2020). Cybersecurity compliance behavior: Exploring the influences of individual decision style and other antecedents. *International Journal of Information Management*, *51*. https://doi.org/10.1016/j.ijinfomgt.2019.102056

Faul, F., Erdfelder, E., Lang, A.-G., & Buchner, A. (2007). G*Power 3: A flexible statistical power analysis program for the social, behavioral, and biomedical sciences. *Behavior Research Methods*, *39*(2), 175–191.

Fornell, C., & Larcker, D. F. (1981). Evaluating structural equation models with unobservable variables and measurement error. *Journal of Marketing Research*, *18*(1), 39–50.

Gregor, S. (2006). The nature of theory in information systems. *MIS Quarterly*, *30*(3), 611–642.

Hair, J. F., Hult, G. T. M., Ringle, C., & Sarstedt, M. (2014). *A primer on partial least squares structural equation modeling (PLS-SEM)*. Thousand Oaks, CA: Sage.

Hair, J. F., Risher, J. J., Sarstedt, M., & Ringle, C. (2019). When to use and how to report the results of PLS-SEM. *European Business Review*, *34*(1), 2–24. https://doi.org/10.1108/EBR-11-2018-0203

Henseler, J., & Dijkstra, T. K. (2018). ADANCO 2.1. Kleve, Germany: Composite Modeling. Retrieved from www.compositemodeling.com

Henseler, J., Hubona, G., & Ray, P. A. (2016). Using PLS path modeling in new technology research: Updated guidelines. *Industrial Management & Data Systems*, *116*(1), 2–20.

Henseler, J., Ringle, C., & Sarstedt, M. (2015). A new criterion for assessing discriminant validity in variance-based structural equation modeling. *Journal of the Academy of Marketing Science*, *43*(1), 115–135.

Herath, T., & Rao, H. R. (2009a). Encouraging information security behaviors: Role of penalties, pressures and perceived effectiveness. *Decision Support Systems*, *47*(2), 154–165.

Herath, T., & Rao, H. R. (2009b). Protection motivation and deterrence: A framework for security policy compliance in organisations. *European Journal of Information Systems*, *18*(2), 106–125. https://doi.org/10.1057/ejis.2009.6

Ifinedo, P. (2014). Information systems security policy compliance: An empirical study of the effects of socialisation, influence, and cognition. *Information & Management*, *51*(1), 69–79.

Johnston, A. C., & Warkentin, M. (2010). Fear appeals and information security behaviors: An empirical study. *MIS Quarterly*, *34*(3), 549–566.

Johnston, A. C., Warkentin, M., & Siponen, M. (2015). An enhanced fear appeal rhetorical framework: Leveraging threats to the human asset through sanctioning rhetoric. *MIS Quarterly*, *39*(1), 113–134.

Kock, N., & Hadaya, P. (2018). Minimum sample size estimation in PLS-SEM: The inverse square root and gamma-exponential methods. *Information Systems Journal*, *28*(1), 227–261. https://doi.org/10.1111/isj.12131

Li, H., Sarathy, R., Zhang, J., & Luo, X. (2014). Exploring the effects of organizational justice, personal ethics and sanction on internet use policy compliance. *Information Systems Journal*, *24*(6), 479–502. https://doi.org/10.1111/isj.12037

Li, H., Zhang, J., & Sarathy, R. (2010). Understanding compliance with internet use policy from the perspective of rational choice theory. *Decision Support Systems*, *48*(4), 635–645. https://doi.org/10.1016/j.dss.2009.12.005

Malhotra, Y., & Galletta, D. (2005). A multidimensional commitment model of volitional systems adoption and usage behavior. *Journal of Management Information Systems*, *22*(1), 117–151.

Myyry, L., Siponen, M., Pahnila, S., Vartiainen, T., & Vance, A. (2009). What levels of moral reasoning and values explain adherence to information security rules? An empirical study. *European Journal of Information Systems*, *18*(2), 126–139.

Nunnally, J. C., & Bernstein, I. H. (1994). *Psychometric theory* (3rd ed.). New York: McGraw-Hill.

Paternoster, R., & Simpson, S. (1996). Sanction threats and appeals to morality: Testing a rational choice model of corporate crime. *Law & Society Review*, *30*(3), 549–584.

Pavlou, P. A., Liang, H., & Xue, Y. (2007). Understanding and mitigating uncertainty in online exchange relationships: A principal-agent perspective. *MIS Quarterly*, *31*(1), 105–136.

Podsakoff, P. M., MacKenzie, S. B., Podsakoff, N. P., & Lee, J.-Y. (2003). Common method biases in behavioral research: A critical review of the literature and recommended remedies. *Journal of Applied Psychology*, *88*(5), 879–903.

Safa, N. S., von Solms, R., & Furnell, S. (2016). Information security policy compliance model in organizations. *Computers & Security*, *56*(C), 70–82.

Siponen, M. T., & Vance, A. (2010). Neutralization: New insights into the problem of employee information systems security policy violations. *MIS Quarterly*, *34*(3), 487–502.

Stone, E. F. (1978). *Research methods in organizational behavior*. Santa Monica, CA: Goodyear.

Wall, J. D., Palvia, P., & Lowry, P. B. (2013). Control-related motivations and information security policy compliance: The role of autonomy and efficacy. *Journal of Information Privacy and Security*, *9*(4), 52–79. https://doi.org/10.1080/15536548.2013.10845690

Webster, J., & Starbuck, W. H. (1988). *Theory building in industrial and organizational psychology*. Chichester, UK: John Wiley and Sons.

Wenzel, M. (2004). The social side of sanctions: personal and social norms as moderators of deterrence. *Law and Human Behavior*, *28*(5), 547–567. https://doi.org/10.1023/B:LAHU.0000046433.57588.71

Whetten, D. A. (1989). What Constitutes a Theoretical Contribution? *Academy of Management Review*, *14*(4), 490–495. https://doi.org/10.5465/amr.1989.4308371

Yazdanmehr, A., & Wang, J. (2016). Employees' information security policy compliance: A norm activation perspective. *Decision Support Systems*, *92*, 36–46.

5

Toward an Exploration of the Scope of Context

5.1 INTRODUCTION

As is typical in many information systems (IS) quantitative behavioural science research (*BSR*) projects, empirical analysis of a new model often results in some of its hypotheses achieving adequate statistical support while others do not, with the latter including some that had achieved statistical support in at least one other previous study, though in some cases there are mixed results. Quantitative *BSR* on a given problem appears to often involve the assumption that there is one 'best' explanatory model that is relevant for all contexts, rather than there could be multiple strong explanatory models, several of which could be appropriate for a given context. Thus, there is often talk about *conflicting results* or absence of *consensus*. However, as noted by Davison and Martinsons (2016), theories developed and tested in one context might not be adequate or valid for another context. Yet the issue of the scope of the context of a new model is often only described in terms of the characteristics of the sample that formed the modelling dataset. We suggest that comparative analysis of the characteristics of the samples that formed the modelling datasets used in different related studies may provide better guidance on the scope of the context of a new model.

5.1.1 Overview on Dimensions of Context

Context (e.g., Whetten, 1989) is typically considered to involve the three non-orthogonal dimensions of *WHO* (i.e., personal characteristics), *WHEN* (i.e., time period), and *WHERE* (i.e., location characteristics). Here we suggest that the *Modelling Environment* is also a dimension of context.

 DOI: 10.1201/9781032678931-5

5.1.1.1 WHO Dimension

The characteristics of the WHO dimension could be considered about the characteristics of the human individual if the unit of analysis is at the level of the individual, or about the characteristics of the organization (or community) if the unit of analysis is at the level of organization (or community). Since the latter set of characteristics is discussed below in the subsubsection on the *WHERE* dimension, here we will focus on characteristics of the human individual that could be relevant.

When the unit of analysis is at the level of the individual, several variables can be considered to be included in the *WHO* dimension (e.g., *Age Group, Gender/Sex, Educational Level, Skill Level, Job Level, Length of Tenure [in current job position], Marital Status, Race/Ethnicity, Big 5 Personality Traits of Openness, Conscientiousness, Agreeableness, Extraversion, and Neuroticism [e.g., Buchanan, Johnson, & Goldberg, 2005], Individual Decision Style, Individual Trust Disposition*). We comment on some of the listed characteristics in Table 5.1.

Several studies have shown that WHO variables can be statistically significant variables. This suggests that *WHO* variables should be considered in the constructing of the requirements of the sample dataset:

- **Gender**: Anwar et al. (2017, p. 5) reported,
  Gender is an important factor mediating human behaviors in general. Our research explores the role of gender in cybersecurity behaviors and beliefs. We compare the constructs of our cybersecurity behavior model between male and female employees in a cross- sectional survey study. The results show that *there are statistically significant gender-wise differences in terms of computer skills, prior experience, cues-to-action, security self-efficacy and self-reported cybersecurity behavior.*
- **Job Level**: Herath and Rao (2009, p. 107) noted,
  There have been some empirical studies that evaluate organisational security practices and their effectiveness; however, the respondents in these studies are typically IT administrators or top-level managers … rather than representatives from the end-user community. The fact that the respondents in prior studies were largely those responsible for setting up and running technical security initiatives raises the question of whether or not their views are likely to be representative of the organisation as a whole.

TABLE 5.1

Comments on Some WHO Variables

Variable	Comments
Age Group	• Typically subjective discretization is issued. • Typically discretized without consideration of whether a given *Age Range* has the same behavioural impacts across different *WHERE*s and *WHEN*s.
Gender/Sex	• Typically based on the traditional biological *Male/Female* binary categorization without consideration of newer categorizations (e.g., LGBTQIA+) advanced by western countries. • It should be noted that for several analytical techniques (e.g., Regression), it is more complicated to deal with categorical variables that involve much more than two possible values.
Educational Level	• Typically raw values are used rather than ranges of values. • Typically used without consideration of whether a given *education level* has the same behavioural impacts across different *WHERE*s and *WHEN*s.
Marital Status	• Typically only involves only binary options (e.g., *Married/ Unmarried*) or a quartet of options (e.g., *Single, Married, Divorced, Widowed*). • Rarely, if ever, considers whether a given option has the same meaning across different *WHERE*s and *WHEN*s.
Individual Decision Style	• Until recently not considered, but Donalds and Osei-Bryson (2020) have shown this to be a statistically significant predictor of cybersecurity compliance behaviour.

5.1.1.2 WHERE Dimension

The *WHERE* dimension can be considered to have several non-orthogonal sub-dimensions including *National Characteristics* and *Organizational Characteristics*.

5.1.1.2.1 WHERE Dimension – National Sub-Dimension

Hofstede (1980) defined a set of cultural dimensions that could impact the behaviours of organizational actors, as outlined in Table 5.2.

Previous research (e.g., Martinsons and Davison, 2007) has reported some of the differences in average societal values of these cultural dimensions for some western and eastern societies, as well as between eastern societies (see Table 5.3).

The characteristics of a given national culture may mean that some behaviors are routine in one national cultural context but may be infeasible in another. Therefore, a sample from a single cultural context may not

TABLE 5.2

Hofstede's Conceptualization of National Culture

Dimension	Description
Power Distance	Reflects the extent to which the members in a society accept the unequal distribution of power.
Individualism–Collectivism	Reflects the degree to which people are able and prefer to achieve an identity and status on their own rather than through group memberships.
Masculinity–Femininity	Reflects the degree to which assertiveness and achievement are valued over nurturing and affiliation.
Uncertainty Avoidance	Reflects discomfort with ambiguity and incomplete information.

TABLE 5.3

Example – Differences in National Cultural Dimensions – East vs. West

Hofstede's Classification Based on Societal Value Score		
	High	**Low**
Power distance	Anglo	Chinese Japanese
Individualism	Anglo	Japanese/Chinese
Uncertainty avoidance	Anglo/Chinese	Japanese
Masculinity	Japanese	
Long-term orientation	Japanese/Chinese	Anglo

Source: Martinsons and Davison (2007, p. 295).

result in sufficient variation in the values of a given variable for it to be determined to be a statistically significant predictor. It is therefore important that cultural issues be taken into consideration when determining the required characteristics of the sample. The extracts below provide examples:

- Gelade, Dobson, and Gilbert (2006, pp. 552–553):
  High employment levels and high rates of economic activity are associated with slightly elevated levels of AC [Affective Commitment] … It might be predicted, however, that more calculative forms of commitment (e.g., Continuance Commitment) would be observed in less economically developed nations, where employees have limited choice in the job market and may be forced to accept unfulfilling work… high levels of cynicism in a society are associated with low

levels of AC ... these results support the proposition that national levels of AC are associated with the general level of positive affect in the population... we might speculate that *Normative Commitment would be high in* **collectivist countries** *and Continuance Commitment high in uncertainty-avoiding and poor countries.*

- Fischer and Mansell (2009, pp. 1351–1353):

  Providing a theoretical account of how Power Distance and Individualism–Collectivism at the culture level can be linked to all three dimensions of Organizational Commitment ... One of the most consistent results was the relatively strong influence of our *macroeconomic control variables* on commitment levels, particularly Affective and Normative Commitment ... These findings suggest that the macroeconomic context has an influence on the *conceptualization of commitment* ... in *Collectivistic settings*, Normative Commitment was found to be higher, which supported our hypothesis. ...The current pattern suggests that individuals in *Collectivistic* and *Power-Distant settings* show higher levels of Continuance Commitment ...the meaning of Continuance Commitment seems to have stronger social implications in Collectivistic and Power-Distant contexts, leading to increased levels of this commitment component.

- Ameen et al. (2021, p. 40):

  The findings of our cross-cultural study challenge existing theories on information security within organizations. Previous studies have reported that the following factors have a significant impact: response cost, self-efficacy, subjective norms, perceived risk vulnerability, response efficacy, perceived severity of sanctions, and perceived certainty of sanctions ... However, *our findings reveal important differences in the significance of these factors among employees in different countries.*

- Chen and Zahedi (2016, p. 204):

  The results support our model and show the divergence between the United States, an exemplar of modern Western society, and China, an exemplar of traditional Eastern society, in forming threat perceptions and in seeking help and avoidance as coping behaviors. Our results also uncovered the significant moderating impacts of espoused culture on the way perceptions of security threats and coping appraisals influence security behaviors. Our findings underline the importance of context-sensitive theory building in security research and provide insights into the motivators and moderators of individuals' online security behaviors in the two nations.

5.1.1.2.2 WHERE Dimension – Organizational Sub-Dimension

The organizational sub-dimension of the WHERE dimension could be described in various ways. We briefly present two of these below while noting that we are not claiming that they are the most representative of the WHERE dimension. For example, as Donalds and Barclay (2022, p. 13) noted in their study,

> The relevance of context and the business environment is also illustrated with the identification of resource-based objectives, such as access to finance, and minimise security costs as factors that can influence the implementation of an organisational InfoSec [information security] strategy. Contextualised in a setting where financial resources are generally limited, such as smaller or less-resourced organisations, the recognition of the relationship of these objectives to Infosec compliance outcomes may be amplified.

Let us consider some possible relationships between an organization's information security culture (*ISC*) and its general organizational culture (*OC*). Some possible relationships could be: (i) "*ISC* is embedded into *OC*"; (ii) "*ISC* is a subculture of *OC*"; and (iii) "*ISC* is separated from *OC*". A review of the "Probable Consequences" column of Table 5.4 suggests that the "*ISC* is embedded into *OC*" relationship may not be feasible in organizations with significant economic constraints. Though this option has other desirable consequences, either the "*ISC* is a subculture of *OC*" or "*ISC* is separated from *OC*" is likely to be more feasible in such organizational contexts, yet the latter has undesirable security compliance consequences. A review of the "Organizational Culture" column would suggest that the values of some of the variables considered in cybersecurity compliance models would be the relationship between the *ISC* and the *OC*, which in turn would be constrained by the economic strength of the organization.

Courtney (2001) presented a set of organizational types, and corresponding organizational decision-making style (see Table 5.5). It seems reasonable to expect that the organizational decision-making style could

TABLE 5.4

Organizational Culture (OC) and Information Security Culture (ISC)

Nature of Relationship	Organizational Culture	Probable Consequences
Type 3: *ISC* is embedded into *OC*	**Information Security Policy:** Created in holistic manner. In addition, there are regular updates on security policy. **Education/Training:** Management makes the awareness program compulsory for all employees. **Budget Practice:** Management allocates budget for security activities annually.	**Security Practices:** Unconsciously become daily routine activities (becoming good security habit). **Investment for security practices:** High cost in implementing security activities.
Type 2: *ISC* is a subculture of *OC*	**Information Security Policy:** Created within IT department and may not have widespread support or knowledge of where they are located. **Education/Training:** Management starts to pay attention to awareness. Employees receive some training on information security. **Budget Practice:** Management acts promptly towards expenses pertaining security activities.	**Security Practices:** Security is employees' routine activities within own dept. **Investment for security activities:** Medium cost in implementing security activities.
Type 1: *ISC* is separated from *OC*	**Information Security Policy:** Created by copying without the means to enforce them. Usually issued by a memo. **Education/Training:** Low awareness. Management does not emphasize security training. **Budget Practice:** Usually part of a budget for IT support.	**Security Practices:** Not a routine activity of employees. **Investment for security activities:** Low cost in implementing security activities.

Source: Lim, Chang, Maynard, &. Ahmad, (2009, p 93).

TABLE 5.5

Courtney's Framework of Organizational Types

Org Type	Learning	Decision-Making Style
Leibnizian	*Creates knowledge by using formal logic and mathematical analysis to make inferences about cause-and-effect relationships … have access only to knowledge generated internally.*	• Formal • Analytical • Bureaucratic
Lockean	*Observe the world, share observations, and create a consensus about what has been observed.*	• Open • Communicative • Consensual
Kantian	*Recognizes that there may be many different perspectives on a problem, or at least many different ways of modeling it. Provided with observations about a decision situation, … chooses the model which best explains the data.*	• Open • Analytical
Hegelian	*Based on the belief that the most effective way to create knowledge is by observing a debate between two diametrically opposed viewpoints about a topic.*	• Conflictual
Singerian	*Seeks the ability to choose the right means for ethical purposes for a broad spectrum of society. … The Singerian inquirer views the world as a holistic system, in which everything is connected to everything else. … Solving complex problems may require knowledge from any source and those knowledgeable in any discipline or profession.*	• Teleological • Cooperative • Ethical

Source: Courtney (2001, pp. 25–28).

impact the level of information systems policy compliance (ISPC). At a minimum, one would expect that given the different perspectives on learning and organizational decision-making, there could be differences in behaviour by actors. For example, in a *Kantian* organization, employees may be more likely to view organizational policies as being of an advisory nature vs. being of an imperative nature, while in a *Leibnizian* organization, the opposite would likely hold.

5.1.1.3 WHEN

Whetten (1989) states that *"theorists should be encouraged to think about whether their theoretical effects vary over time, either because other time-dependent variables are important or because the theoretical effect is unstable for some reason".* For example, Lallie et al. (2021) in the conclusion section of their study noted that

> The COVID-19 pandemic, and the increased rate of cyber-attacks it has invoked have wider implications, which stretch beyond the targets of such attacks. Changes to working practises and socialization, mean people are now spending increased periods of time online. In addition to this, rates of unemployment have also increased, meaning more people are sitting at home online- it is likely that some of these people will turn to cyber-crime to support themselves.

This suggests that the variables associated with such behaviours, practices, and socializations and their relationships among themselves and also with other variables may be different during COVID-19 and on different sides of this pandemic. Similarly, this could apply to other significant events, such as the 2008–2009 financial crisis and the release of the *ChatGPT* family of AI-based data processors.

5.1.1.4 Modeling 'Environment'

The modelling environment involves a set of variables and the set of hypothesized causal relationships included in the model. It should be noted that whether a given hypothesis is supported in a given study could be determined by the presence of other variables and hypothesized causal links (e.g., Balozian, Leidner, & Warkentin, 2017; Siponen & Vance, 2010) in the given model. For example, with respect to our case study (see Chapter 4), *in a post-hoc analysis, we explored if removing IS Security Attitude from the model would result in the Punishment Severity to ISP Compliance Intention link being statistically significant; the result is that this link was then found to be statistically significant. It should be noted that this situation where the addition of a particular construct to a model results in the newly included construct being a statistically significant antecedent of a given target variable, but with some of the other previously significant antecedents no longer achieving statistical significance, has been previously reported in the IS security literature (e.g., Balozian et al., 2017; Siponen & Vance, 2010).*

5.1.2 A Procedure for Planning and Exploring Context

Given a research model that is being considered, we present our procedure for considering its context. This procedure has two phases: *Planning* and *Exploration*.

Phase	Step	Description
PLANNING	1	**Define Desired Sample Characteristics:** Review the extant literature to identify the variables whose variation in values could be impacted by the *WHO*, and/or *WHERE*, and/or *WHEN* of context. Record what is learnt in as much detail as will be useful to determine the desired characteristics of an ideal sample dataset. With respect to details to be recorded, recall the material presented in Section 5.1. Define the required characteristics of an ideal sample dataset for the research model that is to be evaluated.

Model Variable	WHO	WHERE	WHEN

Phase	Step	Description
	2	**Identify Sampling Opportunities for Desired Sample Dataset:** Identify sampling opportunities that are likely to result in the realization of an ideal sample dataset.
	3	**Determine Appropriate Discretization(s):** For each categorical variable (e.g., Age Group, Educational Level), deeply consider what could be an appropriate discretization rather than just using raw values or adopting other commonly used discretizations, particularly if such discretizations have typically resulted in the given variable not shown to be a statistically significant predictor though there are good reasons to believe that it should be. Of course, strong logical arguments should be used to support the discretization that is chosen, such as whether the given discretization has been shown elsewhere to represent differentiating behaviour.

(*Continued*)

Phase	Step	Description
EXPLORATION	4	**Sample Characteristics: Assess Possible Implications of Actual vs. Desired:** After sampling has been done, for each *WHO, WHERE,* and *WHEN* variables, examine its sample metadata (e.g., minimum, maximum, mean, mode, standard variation) to assess whether it has an acceptable level of variation including that for range and variance. One value of doing this step is that it may allow for insight as to the reason why a variable that was hypothesized as having a statistically significant impact was not shown to have such impact in the given study.
	5	**Assess Hypotheses Supported in Previous Studies but not the Given Study:** a) *Impacts of the Modeling Environment:* For each pair of hypotheses of the given causal model that involve the same consequent variable (e.g., Z in $(X \rightarrow Z)$ & $(Y \rightarrow Z)$), note that this pair of antecedents (e.g., X, Y) could be competitors with respect to which is shown to be statistically significant. If one of these relationships (say $(X \rightarrow Z)$) is supported in a given study while the other (say $(Y \rightarrow Z)$) is not supported, though it was supported in other studies, this knowledge may be useful in understanding why the latter was not supported. Data analysis that involves excluding the supported relationship could be used to support that explanation. b) *Impacts of the WHO, WHERE, and WHEN Environments:* For each hypothesis (say $(X \rightarrow Z)$) that was supported in one or more previous studies, but not supported in a given study, determine if there is (are) any *WHO, WHERE,* or *WHEN* contextual difference(s) that could explain the difference in the results (e.g., Figure 5.1).
	6	**Explore Range of Scope of the Context:** Determine other local contexts for which the range of values of the statistically significant direct or indirect predictors can be shown to be similar to that of the sample dataset. This determination could be based on the result of previous studies, or some other means.

5.1.3 Application of the Procedure for Planning and Exploring Context

As far as possible, we will apply the procedure presented in the previous subsections to the case study presented in earlier chapters (see Chapters 3 and 4).

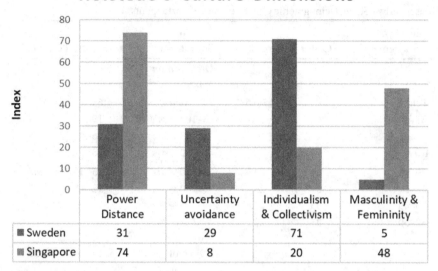

FIGURE 5.1
Example – Differences in Cultural Context.

Source: Björck and Jiang (2006, p. 23).

5.1.3.1 Define Desired Sample Characteristics

The initial desire for the case study was for a sample that involved participants from multiple organizations of different types (including culture)
of a small island developing state (SIDS), where each such organization
had a formal cybersecurity policy for which employees were expected to
be compliant. The expectation is that this would supply adequate variation in values for each of the *WHO* and *WHERE* variables. Consideration
of the WHEN dimension was not included in the scope of the context.
The reader may note that consideration of the possible impact(s) of the
Modeling Environment dimension is considered in Sub-Step 5a.

5.1.3.2 Determine Appropriate Discretization(s)

Table 5.6 displays the metadata on the control variables of the case study.
The reader may note that either no discretization/grouping was done
(e.g., *Job Level, Education Level*) or commonly used discretizations were
used, such as for *Age Range*, were applied. The resulting statistical analysis
showed that "*none of the variables has a significant effect at this stage of the
model testing*". It may well be that for each of these variables, a more appropriate discretization that offers options that are contextually meaningful

TABLE 5.6

Case Study – Socio-Demographic Characteristics of Participants

Category/ Subcategory	Count	Percent (%)	Category/ Subcategory	Count	Percent (%)
Sex:			*Organization tenure:*		
Female	94	60.6	0–5 years	65	41.9
Male	61	39.4	6–10 years	19	12.3
			11–15 years	26	16.8
Age range:			16–20 years	18	11.6
20–29	46	29.7	21–30 years	15	9.7
30–39	39	25.2	31–40 years	12	7.7
40–49	35	22.6			
50–59	35	22.6	*Job level:*		
			Assistant Director	20	12.9
			Head of Department	3	1.9
Education level:			Director	15	9.7
High school degree	11	7.1	Division Head and above	0	0.0
Undergraduate degree	58	37.4	Senior Director	2	1.3
			Line Staff/		
Graduate degree	76	49.0	Entry Level	74	47.7
Other	10	6.5	Supervisor	35	22.6

in a manner that has been shown elsewhere to represent differentiating behaviour might have resulted in a different result with respect to statistically significant impacts. The reader may note that techniques such as decision tree induction (e.g., Osei-Bryson & Ngwenyama, 2014) could be used on pilot study data as part of the process of generating appropriate semantically meaningful discretizations.

5.1.3.3 Identify Sampling Opportunities for Desired Sample Dataset

It was discovered that during the relevant data collection period, opportunities were limited due to inadequate availability of willing participants who worked in a range of relevant organizations. We therefore chose to focus on a single bureaucratic public sector organization whose leadership was interested in participating in a formal study on cybersecurity policy compliance.

5.1.3.4 Actual Sample Characteristics: Assess Possible Implications of Actual vs. Desired

One possible explanation for the lack of observed effects of *IS Security Self-Efficacy* on *IS Security Attitude* pertains to our research setting. We surveyed employees employed by a public sector organisation engaged in *highly procedural and bureaucratic work*. The fact that the study's participants self-reported as having a moderate-high level of competence in performing security related actions (i.e. average *IS Security Self-Efficacy* is 3.7, on a five-point scale), might also be one reason why there was no observed effects. This argument is consistent with and parallels a finding of Dinev, Jahyun, Qing, and Kichan (2009) who reported that *Perceived Ease of Use* of the technology had a significant role in forming a positive attitude towards it for their South Korean respondents but not for their US respondents, who had higher self-reported computer skills (i.e., self-efficacy). Further, it is likely that the highly procedural and bureaucratic nature of the work nullified the effect of the employees' *IS Security Self-Efficacy* on their *IS Security Attitude* (see Chapter 4, Section 4.7.1).

5.1.3.5 Assess Hypotheses Supported in Previous Studies but not the Given Study

5.1.3.5.1 Sub-Step 5a: Impacts of the Modeling Environment Dimension (see Chapter 4)

- … we conducted a post-hoc analysis to explore whether removing *IS Security Attitude* from the model would result in the *General Security Awareness* to *ISP Compliance Intention* link being statistically significant. The link between *General Security Awareness* and *ISP Compliance Intention* was then found to be statistically significant. Thus, the difference in result could be explained by the WHATs and HOWs since *IS Security Attitude* was found to be a fully mediating variable for *General Security Awareness* and other variables in this study.
- Unlike our study, the studies in which this hypothesis achieved adequate statistical support either involved participants from multiple types of organisations and/or did not include the *IS Security Attitude* construct. Further, Balozian et al. (2017, p. 5) observed that *'the severity and certainty of sanctions tested alone are significant; however, when the whole model is tested together, the coercive approach*

loses its significance'. Some possible reasons for the difference in the results include differences in the causal models, the *WHAT*s (such as the presence or absence of the IS *Security Attitude* construct) and *HOW*s (the set of hypotheses) (see Chapter 4).

5.1.3.5.2 Sub-Step 5b: Impacts of the WHO, WHERE, and WHEN dimensions

- In other studies where *IS Security Self-Efficacy* was statistically significant (e.g., Herath & Rao, 2009; Johnston & Warkentin, 2010), the participants were more diverse, consisting of students, faculty and university and business employees. Moreover, the demographic factors of the above cited studies varied more than in this study. This suggests that the difference in the result could be due to the WHERE dimension (e.g., organizational culture of context). According to Wall, Palvia, and Lowry (2013, p. 70), *"it may be that for certain populations, IS Security Self-Efficacy does not have a direct effect on Compliance Intentions"*, supporting our suggestion that the WHERE dimension may account for *IS Security Self-Efficacy's* lack of statistical significant direct impact on *ISP Compliance Intention* in this study" (see Chapter 4).
- Both this study and that of Donalds and Osei-Bryson (2020) involved study participants who were located in the same country. However, while the sample of Donalds and Osei-Bryson (2020) study involved individuals who were employed in a variety of industries, the participants of this study were employed by a bureaucratic public sector organisation. This suggests that the difference in the result could be due to the WHERE dimension of context, but not in the sense of national culture but of organisational culture" (see Chapter 4).

5.1.3.6 Explore Range of Scope of the Context

As discussed in Chapter 4, both the illustrative case study of this investigation and that of Donalds and Osei-Bryson (2020) involved study participants who were located in the same *SIDS*. However, while the sample of Donalds and Osei-Bryson's (2020) study involved individuals who were employed in a variety of industries, the participants of our illustrative case study were employed to a single bureaucratic public sector organization in a *SIDS*. This suggests that the difference in the results could be due to the *Organizational* sub-dimension of the *WHERE* dimension. Thus, it would be reasonable to conjecture that the subset of the causal links of the model that were supported by the data analysis of our case study may be valid for all bureaucratic or other public sector organizations in *SIDS*s, and may also be valid beyond locations in *SIDS*s such as bureaucratic organizations in all countries. However, it does not appear reasonable to conjecture that the results of our illustrative case study data analysis would be valid for all organizational types.

5.1.4 Conclusion

In this chapter, we challenge the prevalent assumption in quantitative BSR studies that a single causal model universally applies to all contexts. This misassumption often leads to researchers reporting "conflicting results" or a "lack of consensus" within a research domain (e.g., Cram, D'Arcy, & Proudfoot, 2019). Instead, we argue that it is more realistic to recognize that causal models are likely valid only within specific contextual subsets. Donalds and Barclay (2022) make that case that scientists should consider the people for whom the expected results are relevant and that there may be different 'best' determinants of employees' ISPC behaviour depending on the context. Unfortunately, the misassumption described above also contributes to insufficient consideration of context both 'before' and 'after' empirical data collection.

To address this critical shortcoming, we introduce a new procedure for rigorously addressing context in the 'pre-data' collection and 'post-data' analysis phases of research. While focusing on the crucial dimensions of *WHO*, *WHERE*, and *WHEN* of context (Whetten, 1989), we also emphasize the significance of the *Modeling Environment* dimension, which is often not considered when attempting to explore the scope of the validity of the theoretical contributions of a given study. Our procedure offers significant practical benefits for researchers, particularly doctoral and graduate students embarking on their empirical projects, as well as for practicing empirical scientists across fields. By promoting a more contextually aware approach to quantitative BSR studies, this chapter aims to improve the validity, replicability, and relevance of research findings. .

REFERENCES

Ameen, N., Tarhini, A., Shah, M. H., Madichie, N., Paul, J., & Choudrie, J. (2021). Keeping customers' data secure: A cross-cultural study of cybersecurity compliance among the Gen-Mobile workforce. *Computers in Human Behavior, 114*. https://doi.org/10.1016/j.chb.2020.106531

Anwar, M., He, W., Ash, I., Yuan, X., Li, L., & Xu, L. (2017). Gender difference and employees' cybersecurity behaviors. *Computers in Human Behavior, 69*(C), 437–443.

Balozian, P., Leidner, D., & Warkentin, M. (2017). Managers' and employees' differing responses to security approaches. *Journal of Computer Information Systems, 59*(3), 197–210. https://doi.org/10.1080/08874417.2017.1318687

Björck, J., & Jiang, K. W. B. (2006). *Information Security and National Culture: Comparison between ERP system security implementations in Singapore and Sweden.* (Master of Science Thesis), Royal Institute of Technology, Stockholm. Retrieved from https://citeseerx.ist.psu.edu/document?repid=rep1&type=pdf&doi=172009e09bc65e3210768f10ae748732f517cbc6

Buchanan, T., Johnson, J. A., & Goldberg, L. R. (2005). Implementing a five-factor personality inventory for use on the internet. *European Journal of Psychological Assessment, 21*(2), 115–127.

Chen, Y., & Zahedi, F. M. (2016). Individuals' internet security perceptions and behaviors: Polycontextual contrasts between the United States and China. *MIS Quarterly, 40*(1), 205–222.

Courtney, J. F. (2001). Decision making and knowledge management in inquiring organizations: Toward a new decision-making paradigm for DSS. *Decision Support Systems, 31*(1), 17–38.

Cram, W. A., D'Arcy, J., & Proudfoot, J. G. (2019). Seeing the forest and the trees: A meta-analysis of the antecedents to information security policy compliance. *MIS Quarterly, 43*(2), 525–554. https://doi.org/10.25300/MISQ/2019/15117

Davison, R. M., & Martinsons, M. G. (2016). Context is king! Considering particularism in research design and reporting. *Journal of Information Technology, 31*(3), 241–249. https://doi.org/10.1057/jit.2015.19

Dinev, T., Jahyun, G., Qing, H., & Kichan, N. (2009). User behaviour towards protective information technologies: The role of national cultural differences. *Inormation Systems Journal, 19*(4), 391–412. https://doi.org/10.1111/j.1365-2575.2007.00289.x

Donalds, C., & Barclay, C. (2022). Beyond technical measures: A value-focused thinking appraisal of strategic drivers in improving information security policy compliance. *European Journal of Information Systems, 1–16.* https://doi.org/10.1080/0960085X.2021.1978344

Donalds, C., & Osei-Bryson, K.-M. (2020). Cybersecurity compliance behavior: Exploring the influences of individual decision style and other antecedents. *International Journal of Information Management, 51.* https://doi.org/10.1016/j.ijinfomgt.2019.102056

Fischer, R., & Mansell, A. (2009). Commitment across cultures: A meta-analytical approach. *Journal of International Business Studies, 40*, 1339–1358.

Gelade, G. A., Dobson, P., & Gilbert, P. (2006). National differences in organizational commitment: Effect of economy, product of personality, or consequence of culture? *Journal of Cross-Cultural Psychology, 37*(5), 542–556.

Herath, T., & Rao, H. R. (2009). Protection motivation and deterrence: A framework for security policy compliance in organisations. *European Journal of Information Systems, 18*(2), 106–125. https://doi.org/10.1057/ejis.2009.6

Hofstede, G. (1980). *Culture's consequences: International differences in work-related values.* London: Sage.

Johnston, A. C., & Warkentin, M. (2010). Fear appeals and information security behaviors: An empirical study. *MIS Quarterly, 34*(3), 549–566.

Lallie, H. S., Shepherd, L. A., Nurse, J. R., Erola, A., Epiphaniou, G., Maple, C., & Bellekens, X. (2021). Cyber security in the age of COVID-19: A timeline and analysis of cyber-crime and cyber-attacks during the pandemic. *Computers & Security, 105*, 102248.

Lim, J., Chang, S., Maynard, S., & Ahmad, A. (2009). "Exploring the Relationship between Organizational Culture and Information Security Culture," in *Proceedings of the 7th Australian Information Security Management Conference*, Perth, Western Australia, pp. 88–97. DOI: 10.4225/75/57b4065130def

Martinsons, M. G., & Davison, R. M. (2007). Strategic decision making and support systems: Comparing American, Japanese and Chinese Management. *Decision Support Systems, 43*(1), 284–300.

Osei-Bryson, K.-M., & Ngwenyama, O. K. (2014). *Advances in Research Methods for Information Systems Research: Data Mining, Data Envelopment Analysis*, Value Focused Thinking: Springer.

Siponen, M. T., & Vance, A. (2010). Neutralization: New insights into the problem of employee information systems security policy violations. *MIS Quarterly, 34*(3), 487–502.

Wall, J. D., Palvia, P., & Lowry, P. B. (2013). Control-related motivations and information security policy compliance: The role of autonomy and efficacy. *Journal of Information Privacy and Security, 9*(4), 52–79. https://doi.org/10.1080/15536548.2013.10845690

Whetten, D. A. (1989). What constitutes a theoretical contribution? *Academy of Management Review, 14*(4), 490–495. https://doi.org/10.5465/amr.1989.4308371

6

Toward a Substantive Exploration of Contributions to Practice

6.1 INTRODUCTION

The expectation that quantitative behavioural science research (*BSR*) studies should involve both rigour and relevance typically requires that they offer actionable guidance for practice. Yet often the guidance for practice that appears in published papers is vague and/or superficial with the reader possibly being left to wonder whether the consideration of such guidance was considered to be an essential part of the research project. Practitioners, while being interested in causal models that have strong predictive power, would also require that the statistically significant predictor variables be actionable. In this chapter, we provide an analysis of how considerations of the range of options offered by the types of predictor variables (i.e., *extrinsic* vs. *intrinsic*) can be used to determine implications for practice that are offered by a causal model while also considering feasibility factors. In this chapter, we present an approach that could be used to expose actionable guidance that is offered by the results of a causal model.

6.2 CONSTRUCTING AND EVALUATING THE STUDY'S CONTRIBUTIONS TO PRACTICE

Now, while a decision option may seem to be available at a given point in time, it may not be feasible in the given organizational context. For instance, a decision on a given intrinsic variable could be infeasible due to

DOI: 10.1201/9781032678931-6

legal constraints (e.g., one involving assignment to a project that factored in gender, race, age), while a decision on a given extrinsic variable could be infeasible due to economic (e.g., too costly), technical (e.g., inconsistent with available technological resources), or other feasibility constraints. Thus, the models that are most relevant for a given context are those that are actionable, have high predictive power, and have high explanatory power (particularly if legal constraints require explanations for each decision) while also satisfying other feasibility constraints.

TELOS-PDM (Burch, 1992) is a framework that has been used to evaluate organizational projects using two sets of criteria (Bryson, Ngwenyama, & Mobolurin, 1994): its expected contribution to a firm's strategic goals of improving *Productivity, Differentiation,* and *Management* (*PDM*); and whether it is feasible in terms of *Technical, Economic, Legal, Operational,* and *Scheduling* (*TELOS*) concerns. With respect to the *PDM* aspect, any *Contributions to Practice* guidance should result in an improvement in Productivity (i.e., relative trade-off between additional outputs compared to additional inputs (e.g., economic cost)) and should involve some novelty (i.e., Differentiation) for it to be considered to be a contribution. It also seems reasonable to expect that implementation of any *Contributions to Practice* guidance has to be feasible with respect to relevant *TELOS* concerns, although not all *TELOS* factors may be relevant to a given guidance.

To develop and analyse *Contributions to Practice*, we propose an analytical framework, as presented in Table 6.1, that is inspired by the PDM-TELOS framework. Our adaptation involves (1) exclusion of the management goal as it is not easily useful for our analytical purpose and (2) contextualization of the description of each of the other elements (i.e., goals and factors) of the original framework. We do not suggest that this framework is the best or the only appropriate one for this purpose, but rather that some analytical framework should be used by authors as they consider how and what to present as their study's *Contributions to Practice*.

6.2.1 Examples of Analytical Framework-Based Evaluation of Some Previous Guidance

Before demonstrating how our analytical framework could be used to both construct and evaluate *Contributions to Practice* guidance

TABLE 6.1

Analytical Framework

Goal	Description
Productivity	• The relative predictive strength of a model that is parsimonious. • Trade-off between additional value vs. additional cost regarding following the guidance, where the additional value is based on the increase in the level of the dependent variable that is caused by raising the levels of one or more input variables as proposed by the guidance; and the cost is based on the increase(s) in the relevant level(s) of the input variable(s).
Differentiation	• The causal model is much different from any previous model. • The guidance is novel as it is much different from previously published guidance.

Feasibility Factor	Description
Technical	Availability of required technical resources (i.e., Hardware, Software, Infrastructure, and Personnel) associated with the guidance.
Economic	Economic cost associated with a given guidance.
Legal	Whether each guidance is in conformance with the given set of Societal Laws and Organizational Regulations.
Operational	Abilities and Attitudes of those who would implement and/or be affected by the guidance.
Scheduling	Time period at which the guidance is feasibly actionable.

associated with a given causal model, we will present and comment on guidance from some previously published papers. The focus will be on the elements of our analytical framework that appear to be most obviously relevant to each guidance, and as such should be considered in the development and articulation of the guidance. The reader may note that in Table 6.2, the *D*ifferentiation column is checked for each guidance in order to highlight the fact that for a guidance to be considered to be a 'contribution' it should possess novelty. Also note that typically not all elements of our analytical framework will be obviously relevant in the assessment of each guidance. With respect to interpreting a check mark, its presence in the table means that the given criterion is relevant to the guidance; it doesn't mean that the criterion was actually covered in the guidance.

TABLE 6.2

Analytical Framework–Based Analysis of Several Previous Guidance

Guidance	P	D	T	E	L	O	S	Comments
Vedadi and Warkentin (2020, p. 444): *"it is important for IT managers and policy makers to predict the IT security phenomena that are highly likely to become popular, thoroughly identify their disadvantageous implications."*	√	√	√	√			√	Although it appears to aim for better *Productivity*, it does not consider the *Technical, Economic,* and *Scheduling* factors.
Siponen, Pahnila, and Mahmood (2010, p. 70): *" … we recommend that managers motivate employees to form an intention to comply with information security policies using nonreward-oriented motivational factors such as setting an example themselves by visibly complying with these policies, providing information security education and training, and clearly and forcefully stating the sanctions for security policy noncompliance."*				√		√		Considers *Operational* feasibility, though *Novelty* is required of each contribution it is lacking here.
Hensel and Kacprzak (2021, p. 231): *" … Finally, we need to once again emphasise the importance of the proper application of sanctions such as those studied in our paper. The managers at the studied organisation used the best available evidence to devise their sanctioning policy. That is, since short periods of cyberloafing might bring beneficial consequences for the employees' well-being and performance, the intent of the exercise was to reduce rather than to eliminate cyberloafing."*				√		√		Considers *Operational* feasibility and appears to be *Novel*.
Chen and Zahedi (2016, p. 219): *"Moreover, individuals' security behaviors play an important role in reducing the success of cybercrimes and increasing the safety of the Internet environment. Therefore, private and public organizations should encourage individuals to adopt coping behaviors that promote safety and counter security threats."*				√				Though *Novelty* is required of each contribution, it is lacking here.

(Continued)

TABLE 6.2 (Continued)

Guidance	P	D	T	E	L	O	S	Comments
Chen and Zahedi (2016, p. 219): *"Moreover, distinct national differences could be utilized when developing strategies to promote protective behaviors."*	√					√		Considers *Operational* feasibility, but though *Novelty* is required of each contribution, it is lacking here.
Li et al. (2019, p. 22): *"Employees often take their cues from management in terms of security compliance. Thus, managers need to lead by example to create and maintain a secure workplace."*						√		Considers *Operational* feasibility, but though *Novelty* is required of each contribution, it is lacking here.
Li et al. (2019, p. 22): *"Management can provide regular in-house information security awareness workshop and training to positively shape the attitudes of their employees regarding information security issues."*	√					√		Considers *Operational* feasibility, but though *Novelty* is required of each contribution, it is lacking here.
Li et al. (2019, p. 22): *"Information and cybersecurity procedures, guidelines, and policies should be written and constructed in a user-friendly manner to encourage non-IT employees to comply with information security matters."*	√	√				√		Considers *Operational* feasibility but is not *Novel*.
Li et al. (2019, p. 22): *"Establishing a supportive organizational environment, such as general exposure to emerging security technologies and relevant incentives to encourage employees to develop and improve the necessary skills and knowledge that are needed to safeguard the company's information assets."*	√		√			√		Considers *Operational* feasibility but is not *Novel*, and does not consider *Economic* feasibility.
Li et al. (2019, p. 22): *"Companies should also note the differences in behavior between male and female employees in order to optimize recruitment strategies for leadership positions."*					√			This may violate *Legal feasibility*.

Yáñez-Valdés and Guerrero (2023, p. 8): "First, the development of integrated policies that support technological initiatives and associative processes between institutions interacting with the platform. …. in emerging economies."	√		These guidance are generic and also not *Novel*.
Rese and Tränkner (2024): "Testing the chatbot before going online and analysing mistakes with chatbot analytic tools can increase chatbot performance … In addition, the managerial implications relate to improvement in the correct use of standard phrases, such as greetings, and the minimal use of sentences to restart the conversation."	√		These guidance are generic and also not *Novel*.
Lowry, Xiao, and Yuan (2023, p. 944): "2P lending platforms should alleviate lenders' loan risks by offering them more soft information on borrowers … including their photos, social capital, and external online links, … PPDai could use its information architecture to disclose certain elements of this information to its members. In this scenario, lenders could use the information to improve their lending decisions."	√ √ √	√	Appears to aim for better *Productivity*. While being *Technically* feasible, this guidance may violate *Legal* feasibility as it could lead to discrimination based on race, class, etc.

6.2.2 An Analytical Process for Construction of Guidance

Here we present our analytical framework-based process showing how it could be used to both construct and evaluate *Contributions to Practice* guidance.

Step	Description
1	Identify each of the variables that is either a *D*irectly or *I*ndirectly statistically significant predictor of the dependent variable(s). Determine the subset of these variables that could be actionable.
2	For each actionable variable identified in Step 1:
	a) Determine whether it is an *E*xtrinsic or *I*ntrinsic variable.
	b) Use the coefficients of the relevant causal links to estimate its total impact on the dependent variable.
3	For each construct identified in Step 1:
	a) Use the analytical framework, to explore its elements that seem highly relevant to affecting the level of the construct.
	b) Also consider other issues (e.g., Hofstede's Cultural Dimensions) that may be relevant.
4	For each mediator variable, do similarly what was done in Steps 1 and 2 for the dependent variable.
5	Use the insights from doing Steps 1–4, to construct an appropriate set of *Contributions to Practice* guidance.

6.3 ILLUSTRATIVE EXAMPLES

In this section, we will focus on demonstrating how our analytical framework could be used to both construct and evaluate *Contributions to Practice* guidance associated with a given causal model. We will use two illustrative studies: the first from the earlier chapters of this book and the second from a well-cited journal article. For each study, we will use our analytical framework to explore the statistically significant predictors of the associated dependent variables, particularly in order to expose elements that are relevant to the construction of useful guidance. With respect to Illustrative Example 2, we will also use our analytical framework to explore the *Contributions to Practice* guidance that were presented in the relevant journal article.

6.3.1 Example 1

This example involves the model previously presented in earlier chapters (e.g., Chapter 4).

6.3.1.1 Considering Productivity

The model presented in Figure 6.1 is used as the basis for guidance to practitioners. This model has one of the highest reported *R*-squared for *ISP Compliance Intention* (i.e., 0.764) and one of the highest for *IS Security Attitude* (i.e., 0.535) in the literature.

6.3.1.2 Considering Differentiation (i.e., Novelty)

The model presented in Figure 6.1 demonstrates that with respect to *ISP Compliance Intention*, high explanatory and predictive power can be achieved by a model that does not involve as its direct statistically significant antecedents variables such as *Punishment Severity, Detection Certainty,* and *Response Cost* and some of the other variables identified in previous studies as statistically significant antecedents. This model also demonstrates that with respect to *IS Security Attitude*, high explanatory and predictive power can be achieved by a model that does not involve as its direct statistically significant antecedents variables such as *IS Security Self-Efficacy* and some

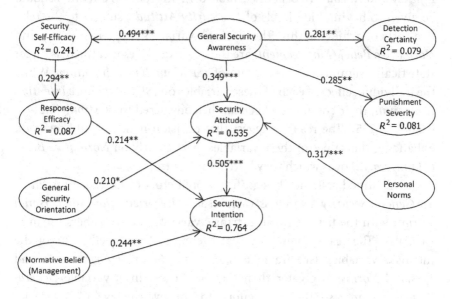

FIGURE 6.1
Causal Model of Example 1.

of the other variables identified in previous studies as statistically significant antecedents.

6.3.1.3 Considering Scheduling Feasibility

The variables in a causal model can be characterized as being extrinsic (X) or intrinsic (N) (see, e.g., Table 6.3). The options that are available to a decision-maker for making an actionable decision would be different for an intrinsic variable versus an extrinsic variable. For example, with respect to the intrinsic variable *General Security Orientation*, actionable options may only be available at the time that the employee is hired or assigned to a project for which ISP compliance is critical. On the other hand, with respect to the extrinsic variable *General Security Awareness*, actionable options (e.g., training) may be available at various times.

6.3.1.4 Considering Economic Feasibility

With respect to *ISP Compliance Intention*, the identified statistically significant antecedents are *IS security Attitude*, *Response Efficacy*, and *Normative Belief (Management)*, with the first two being mediators and the other an independent extrinsic variable. The reader may note that the impact of *IS security Attitude* is more than twice the sum of the impacts of *Response Efficacy* and *Normative Belief* (see Table 6.4). Thus, much attention should be given to having high levels of *IS security Attitude*, subject to the relative costs of achieving this. The extrinsic variables *Response Efficacy* and *Normative Belief (Management)* are the other variables identified as direct statistically significant antecedents of *ISP Compliance Intention*. While the individual effect of each of these variables on *ISP Compliance Intention* is *medium*, their combined effect can be considered to be *strong* (0.214 + 0.244 > 0.235). The results suggest that decision-makers could focus on enhancing the values of these variables, particularly on *Normative Belief (Management)* over which they have more control.

The identified statistically significant predictors of *IS security Attitude* are *General Security Awareness*, *General Security Orientation*, and *Personal Norms*, with the first being an extrinsic variable and the others intrinsic variables. The reader may note that the sum of the coefficients of the intrinsic variables (see Table 6.3), *General Security Orientation* and *Personal Norms*, is greater than that of the extrinsic variable *General Security Awareness*. Therefore, efforts to improve the level of *IS security Attitude* should involve assessment of the employee's current levels of

TABLE 6.3

Statistically Significant Antecedents of Security Attitude

Variable	X/N	S/I	Impact	P	D	T	E	L	O	S	Comments
General Security Awareness	X	S	0.349						√		
Personal Norms	N	S	0.317						√	√	Assessment of employees' *Personal Norms* should be made at the time of hire, or before assignment to security-sensitive projects.
General Security Orientation	N	S	0.210						√	√	Assessment of employees' *General Security Orientation* should be made at the time of hire, or before assignment to security-sensitive projects.

X: Extrinsic variable; *N*: Intrinsic variable; *S*: Straight (Direct) Impact; *I*: Indirect Impact.

TABLE 6.4

Statistically Significant Antecedents of Security Policy Compliance Intention

Variable	X/N	S/I	Impact	D	M	T	E	L	O	S	Comments
Security Attitude	X	S	0.505						√		
Normative Belief	X	S	0.244						√		
Response Efficacy	X	S	0.214						√	√	
General Security Awareness	X	I	0.207								
Personal Norms	N	I	0.160								
General Security Orientation	N	I	0.106								
Security Self-Efficacy	X	I	0.063								

X: Extrinsic variable; *N*: Intrinsic variable; *S*: Straight (Direct) Impact; *I*: Indirect Impact.

General Security Orientation and *Personal Norms* at the time of his/her selection for membership on security-sensitive teams while also having training to improve his/her level of *General Security Awareness*. If this is not done, then efforts to improve *IS security Attitude* could incur additional costs to improve the level of *General Security Awareness* and yet yield sub-optimal results.

6.3.2 Example 2

This involves the model presented in Herath and Rao (2009) (Figure 6.2).

Table 6.6 is derived from Table 6.5 but focuses on the total impacts of two subsets of the predictor variables: the technologically oriented predictors (i.e., *Detection Certainty* and *Resource Availability*) and the other predictors (i.e., *Subjective Norm, Organizational Commitment, Self-Efficacy,* and *Descriptive Norm*). A review of the data in Table 6.6 indicates that total impact of the technologically oriented predictors is less than 1/3 (0.242 vs. 0.791) of the total impact of the other predictors. This suggests that unless their levels are already high, more emphasis should be placed on enhancing the levels of the predictors that do not necessarily involve the *Technical feasibility* factor. This further suggests that it could be useful for the organization to assess new employees at the time of hire with respect to these predictors (i.e., *Subjective Norm, Organizational Commitment, Self-Efficacy,*

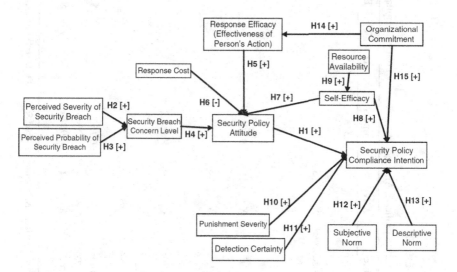

FIGURE 6.2
Causal Model of Example 2.

TABLE 6.5

Statistically Significant Antecedents of Security Policy Compliance Intention

Variable	X/N	S/I	Impact	P	D	T	E	L	O	S	Comments
Subjective Norm		S	0.313							√	
Organizational Commitment	X	S	0.202						√	√	The *Individualism* dimension of national culture (Hofstede, 1980) may be relevant. It is possible that employees of *collectivist* cultures may have a higher tendency to *Organizational Commitment* than those of *individualistic* cultures (Fischer & Mansell, 2009; Gelade, Dobson, & Gilbert, 2006).
Descriptive Norm	X	S	0.101						√		
Gender	N	S	0.098					√			Using *Gender* as the basis to assign employees to projects in which *Security Policy Compliance* is critical would likely violate the *Legal* feasibility factor.
Self-Efficacy	X	S	0.175			√			√		Since *Resource Availability* does impact this predictor, given the magnitudes of the corresponding R-squared and coefficient values, it seems reasonable to infer that some employees could have a high level of *Self-Efficacy* even when there is low *Resource Availability*.
Detection Certainty	X	S	0.155			√	√				Impacting *Detection Certainty* most likely would involve *Technical* considerations, and as a consequence *Economic* considerations.
Resource Availability	X	I	0.087		√		√			√	

X: Extrinsic variable; *N*: Intrinsic variable; *S*: Straight (Direct) Impact; *I*: Indirect Impact.

and *Descriptive Norm*), and also the same for current employees. It is possible that there could be clusters of employees that are at different levels with respect to these predictors, and if such clusters were identified, then targeted approaches could be used for appropriate enhancements of the levels of these predictors.

As stated earlier, we will now use our analytical framework to explore the *Contributions to Practice* guidance that were presented in Herath and Rao (2009). In the *Comments* column of Table 6.7, we illustrate how our Analytical Framework-Based Process could be used to enhance the guidance presented in the relevant journal article.

TABLE 6.6

Statistically Significant Antecedents and Technology Involvement

Technology factor is always involved			Technology factor is not always involved		
Predictor of Intention	**S/I**	**Impact**	**Predictor of Intention**	**S/I**	**Impact**
Detection Certainty	S	0.155	*Subjective Norm*	S	0.313
Resource Availability	I	0.087	*Organizational Commitment*	S	0.202
Total Impact		**0.242**	*Self-Efficacy*	S	0.175
			Descriptive Norm	S	0.101
			Total Impact		**0.791**

X: Extrinsic variable; *N*: Intrinsic variable; *S*: Straight (Direct) Impact; *I*: Indirect Impact.

TABLE 6.7

Analytical Framework-Based Comments on Herath & Rao's (2009) Guidance

Herath and Rao (2009) Study's Guidance	Comments
"…*it is necessary for IT management to communicate the reality of security threats to organisational end-users*"	This is a generic comment that is not novel.
"*Resource availability was found to significantly enhance employees' abilities to perform the necessary security-related actions. … As such, managers need to make security policy-related resources easily available to employees*"	Our analysis above suggests that investing in some of the other variables (e.g., *Subjective Norm* and *Organizational Commitment*) would be more beneficial and should be given higher priority.

(Continued)

TABLE 6.7 (Continued)

Herath and Rao (2009) Study's Guidance	Comments
"… *(self-efficacy) were found to have an effect on both security policy attitudes and intentions to comply with policies. Thus, employee self- efficacy is likely to result in favourable attitudes and more compliance intentions*"	Given the *Resource Availability* only explains 25% of the variance of *Self-Efficacy*, then some employees might have high levels of *Self-Efficacy* even in the absence of *Resource Availability*, while for others this is not the case. It may thus be useful and *Economically* feasible to have a targeted approach to increasing the level of *Resource Availability* in order to raise the level of *Self-Efficacy*, in order to raise the level of *Compliance Intention*.
"*Managers can improve security compliance by enhancing the security climate in their organisation*"	This guidance relates to *Subjective Norm*. Given the high potential impact and likely relatively low *Economic* cost of impacting this variable, high priority should be given to impacting this variable.
"*Employee organisational commitment was found to have a significant impact on both the policy compliance intentions and perceived effectiveness of employee actions. The extant literature in organisational behaviour … can give us insights into the managerial actions that promote employee involvement*"	It might be useful to assess at the time of hire the tendency to *Organizational Commitment*. Earlier we suggested that the *Individualism* dimension of national culture (Hofstede, 1980) may be relevant to this tendency.

6.4 CONCLUSION

This chapter highlights the critical need for quantitative BSR to balance rigor with relevance, ultimately providing actionable guidance for practitioners. While robust causal modeling is essential, its value diminishes if the identified statistically significant predictors cannot be readily translated into practice. This chapter addressed the gap between the desire for actionable guidance from causal models and the often vague or superficial recommendations found in research publications. We address this gap by presenting an approach that systematically analyses predictor variables (extrinsic vs. intrinsic) within a causal model, extracting actionable implications while considering feasibility factors.

To demonstrate the practical application of our framework, we presented two illustrative examples, showcasing how the process can be used to extract actionable guidance from different causal models.

Our analytical framework and an associated process ensure that research findings are not merely theoretical but offer tangible recommendations for practitioners. By emphasizing the translation of statistically significant results into actionable guidance, this chapter bridges the divide between research and practice. It empowers practitioners to implement evidence-based strategies informed by robust BSR and drive meaningful change in their fields.

REFERENCES

Bryson, N., Ngwenyama, O. K., & Mobolurin, A. (1994). A qualitative discriminant process for scoring and ranking in group support systems. *Information Processing & Management, 30*(3), 389–405.

Burch, J. (1992). *Systems analysis, design and implementation.* Boston: Boyd and Fraser.

Chen, Y., & Zahedi, F. M. (2016). Individuals' internet security perceptions and behaviors: Polycontextual contrasts between the United States and China. *MIS Quarterly, 40*(1), 205–222.

Fischer, R., & Mansell, A. (2009). Commitment across cultures: A meta-analytical approach. *Journal of International Business Studies, 40,* 1339–1358.

Gelade, G. A., Dobson, P., & Gilbert, P. (2006). National differences in organizational commitment: Effect of economy, product of personality, or consequence of culture? *Journal of Cross-Cultural Psychology, 37*(5), 542–556.

Hensel, P. G., & Kacprzak, A. (2021). Curbing cyberloafing: Studying general and specific deterrence effects with field evidence. *European Journal of Information Systems, 30*(2), 219–235.

Herath, T., & Rao, H. R. (2009). Protection motivation and deterrence: A framework for security policy compliance in organisations. *European Journal of Information Systems, 18*(2), 106–125. https://doi.org/10.1057/ejis.2009.6

Hofstede, G. (1980). *Culture's consequences: International differences in work-related values.* London: Sage.

Li, L., He, W., Xu, L., Ash, I., Anwar, M., & Yuan, X. (2019). Investigating the impact of cybersecurity policy awareness on employees' cybersecurity behavior. *International Journal of Information Management, 45,* 13–24. https://doi.org/10.1016/j.ijinfomgt.2018.10.017

Lowry, P. B., Xiao, J., & Yuan, J. (2023). How lending experience and borrower credit influence rational herding behavior in peer-to-peer microloan platform markets. *Journal of Management Information Systems, 40*(3), 914–952. https://doi.org/10.1080/074212 22.2023.2229128

Rese, A., & Tränkner, P. (2024). Perceived conversational ability of task-based chatbots–Which conversational elements influence the success of text-based dialogues? *International Journal of Information Management, 74,* 102699. https://doi.org/10.1016/j.ijinfomgt.2023.102699

Siponen, M. T., Pahnila, S., & Mahmood, M. A. (2010). Compliance with information security policies: An empirical investigation. *Computer, 43*(2), 64–71. https://doi.org/10.1109/MC.2010.35

Vedadi, A., & Warkentin, M. (2020). Can secure behaviors be contagious? A two-stage investigation of the influence of herd behavior on security decisions. *Journal of the Association for Information Systems, 21*(2). https://aisel.aisnet.org/jais/vol21/iss2/3

Yáñez-Valdés, C., & Guerrero, M. (2023). Assessing the organizational and ecosystem factors driving the impact of transformative FinTech platforms in emerging economies. *International Journal of Information Management, 73*, 102689. https://doi.org/10.1016/j.ijinfomgt.2023.102689

7

Towards Making the Results of Likert-Scale Behavioural Science Research Actionable in Organizational Settings

7.1 INTRODUCTION

Information systems (IS) research can typically be categorized as involving either the design science or behavioural science methodologies, with the latter being dominant in terms of published papers in what are considered to be the leading IS research journals. Those projects involving the behavioural science methodology can further be sub-categorized as being quantitative or qualitative, with the former dominating. In this chapter, we are concerned with quantitative behavioural science research projects whose outputs are causal theoretical models that both explain and predict (Gregor, 2006), and in which the items associated with the constructs were measured on a Likert scale. Although the majority of the quantitative behavioural science research projects attempt only to provide explanation, there are examples of some that both explain and predict. For example, Gregor (2006, p. 628) noted that the *'Technology Acceptance Model (TAM) (Davis et al., 1989) and DeLone and McLean's dynamic model of information success (1992, 2003) both aim to explain and predict.'*

In this chapter, we assume that such causal models would involve statistically significant coefficients for relevant links, an assumption that holds for a very large subset of published behavioural science research projects. So, we are concerned with addressing how such quantitative causal models with high explanatory power and high predictive power could be made actionable for decision-making by practitioners. We take the position that if in such a scenario the items scores are based on a Likert scale then the factor scores can also be calculated on the given Likert scale, and any

DOI: 10.1201/9781032678931-7

assessment of the current and future states of the organization in terms of independent, mediator and dependent factors should be done in terms of the given Likert scale. In this chapter, we present a new method that involves the use of concepts from mathematical programming and multi-criteria decision analysis (MCDA) to make the quantitative causal model actionable for decision-making. We also provide an illustrative example.

7.2 DESCRIPTION OF THE PROBLEM

Given a quantitative causal model with high explanatory and predictive power, and statistically significant coefficients for relevant links:

- the current Likert-scale level for each predictor (e.g., independent variable, mediator variable) and the dependent variable have been determined;
- for each predictor variable, the estimated cost of moving to each higher level can be estimated;
- for the dependent variable, the value of the current level and higher levels can be estimated;
- there is a specified total budgeted amount (say B) that is available for upgrading the levels of the predictor variables (e.g., *System Quality, Utilization, Ease of Use*) in order to maximize the difference between the resulting value of the dependent variable (e.g., *User Performance*) and the total cost of upgrading the levels of the predictor variables.

Addressing this problem is important if the results of behavioural science research are to be actionable by practitioners. The discussion or conclusion sections of behavioural science research journal papers typically include a subsection that focuses on implications of the results for research and for practitioners. However, in the real-world situations where the decision-maker has to make decisions that offer the most value in the context of limited budgetary and other resources, the typical recommendations do not factor in such context. For example, (Pavlou, 2003, pp. 93–94) in his Implications for Practice subsection states:

> The study has important practical implications for influencing on-line consumer purchasing behavior. Web retailers should acknowl-edge that consumer trust and risk constitute a tremendous barrier

to on-line transactions. ... Web retailers could employ several trust-building mechanisms to manipulate favorable consumer attitudes and ultimate transaction behavior.

More recently Ghasemaghaei, Hassanein, and Turel (2016, p. 103) offered the following:

From a practical perspective, our findings clearly show that ..., firms should employ thorough selection processes when acquiring data analytics tools to ensure that the selected tools will most closely match their data, people and tasks. They could also redesign their tasks to take better advantage of available analytical tools.

It can be easily seen from the above that though the recommended actions would have cost implications for the given organization and hopefully also verifiable benefits, such concerns do not appear in the Implications for Practice subsection.

For several years, there has been the *Rigor* vs. *Relevance* debate. According to Gulati (2007, p. 775), 'A *long-standing debate among management scholars concerns the rigor, or methodological soundness, of our research versus its relevance to managers.*' Many academic behavioural science researchers have rejected this as being an *Either/Or* situation and have attempted to address relevant problems using rigorous methods. However, possibly because much of academic preparation and practice of behavioural science researchers is focused on the 'explanatory' type of theory (Gregor, 2006), there has been inadequate concern about how the results of such research could be used by decision-makers beyond the 'ritualistic' requirement of including an 'Implications for Practice' subsection in papers and dissertations. The research problem addressed in this chapter is aimed at making the results of research of the 'explanatory and predictive' type of theory relevant to such decision-makers.

7.3 OVERVIEW OF THE ANALYTIC HIERARCHY PROCESS (AHP)

The analytic hierarchy process (AHP; Saaty, 1980) is aimed at facilitating decision-making in problems that involve multiple criteria. Its use involves first structuring the set of criteria into a hierarchy, generating weights for the subset of the criteria at each level of the hierarchy, for each leaf-level criterion determining the relative weight of each alternative with respect to this criterion, and finally synthesizing the previously generated sets of weights to produce an overall relative score for each alternative.

Weight generation is done implicitly via the elicitation preference information from the decision-maker and to produce a corresponding priority vector ($\mathbf{w}$). The preference information is represented numerically using a positive reciprocal matrix $A = \{a_{ij}\}$ with $a_{ij} = 1/a_{ji}$, where a_{ij} is a rational number that is the numerical equivalent of the comparison between objects 'i' and 'j'. The priority vector $\mathbf{w}$ may then be obtained from the comparison matrix A using a variety of techniques.

Traditional approaches for using the AHP in the group decision-making context require that the comparison information with regard to each pair of objects be determined by either (1) consensus vote or (2) the geometric averaging of the preference information provided by the individual group members. In the first option, the entire group provides a single numeric value for each pair of objects, resulting in a 'consensus' matrix $A^{GM} = \{a_{ij}\}$. In the second option, each individual (say 't' in T, T being the index set of the group members) provides a numeric value a^t_{ij} that reflected her/his view of the relative importance of object 'i' compared to object 'j'. The 'consensus' pairwise comparison matrix $A^{GM} = \{a_{ij}\}$ is computed by the formula $a_{ij} = (\Pi_{(t \,\epsilon\, T)} \, a^t_{ij})^{1/M}$, where '$M$' is the number of group members.

7.4 DESCRIPTION OF THE PROPOSED SOLUTION

7.4.1 Mathematical Programming Formulation

Let the relevant Likert scale involve the set of integers from 1 to j_{Max}. Let $f_{(i)} = j_{Min(i)}$ be the current level of variable 'i' and $J_{(i)}$ be the set of integers from $j_{Min(i)}$ to j_{Max}. Let I be the set of decision variables; $c_{i(f,j)}$ be the cost of upgrading the f variable 'i' from Likert-scale level $j_{Min(i)}$ to level j; and B be the total available budgeted amount for upgrading the values of all decision variables.

1a: $\sum_{j \,\epsilon\, Jf} x_{ij} = 1$ $\qquad\qquad$ $\forall \, i \in I.$

1b: $x_{ij} \in \{0, 1\}$ $\qquad\qquad$ $\forall \, i \in I, j \in J_{(i)}.$

1c: $\sum_{j \in If} j x_{ij} - z_i = 0$ $\qquad\qquad \forall i \in I.$

1d: $\sum_{j \in I} f c_{i(f,j)} x_{ij} - b_i = 0$ $\qquad \forall i \in I.$

1e: $\sum_{i \in I} b_i \leq B$

Constraints 1a and 1b ensure that only a single $x_{ij} = 1$ for a given $i \in I$. Constraint 1c ensures that z_i describes the Likert-scale level that corresponds to the j for which $x_{ij} = 1$ for $i \in I$. Constraints 1d and 1e ensure that the upgrade costs do not exceed the budgeted amount.

Assuming that the equation provided by the regression analysis that describes the relationship between decision variables and the output is:

$$Y = b_0 + \sum_{i \in I} b_i X_i.$$

Given constraints 1a–1e above, this equation could be expressed as

$$Y = b_0 + \sum_{i \in I} b_i z_i.$$

So our problem becomes: $Y_{Max} = Max \, (\beta_0 + \Sigma_{i \in I} \beta_i z_i) \mid$ constraints 1a–1e.

It should be noted that although Y_{Max} might not be an integer value we can apply a rounding function such that $Y_{MaxRnd} = Rnd(Y_{Max}) \in \{1, 2, 3, \ldots, 7\}$ is on the Likert scale. Now having the level of the output variable, we can then identify the benefit value v_{MaxRnd} associated with the level YMaxRnd.

7.4.2 Estimating Cost and Benefit Values

Our method proposes the use of pairwise comparison (PC) approaches (e.g., Bryson, 1996; Saaty, 1980) to estimate both cost and benefit values based on information elicited from a group of domain experts. PC approaches have been previously used to estimate costs. For example, An, Kim, and Kang (2007) used it to estimate costs in construction projects and found it to offer greater accuracy than the two other methods that they considered.

Given $f_{(i)}$, the current level of predictor variable 'i', then we would be interested in estimating the costs of upgrading predictor variable 'i' from its current level to each higher level (i.e., $(f_{(i)} + 1)$, $(f_{(i)} + 2)$, ..., j_{Max}). It is reasonable to expect that each of these higher levels, level $(f_{(i)} + 1)$, would be the easiest for the domain experts to directly estimate its upgrade cost.

For higher levels, direct estimation may be more challenging, and so relative estimation may be more appropriate.

Let b_{pq} be the numeric ratio that represents the upgrade cost for upgrading from level $f_{(i)}$ of predictor variable 'i' to its level 'p' relative to the corresponding upgrade cost for upgrading its level 'q'. If b_{pq} is elicited for each integer pair (p, q) where $p \in [(f_{(i)} + 2), j_{Max}]$ and $q \in [(f_{(i)} + 1), (j_{Max} - 1)]$, then weight vector generation methods (e.g., Bryson, 1995; Bryson & Joseph, 1999; Saaty, 1980) can be used to generate the corresponding weight vector. It should be noted that these methods can be applied in both the individual and group decision-making contexts. It should also be noted that there are weight vector generation methods that can be appropriate even in situations where there are some (p, q) pairs for which PC data was not provided.

Let $\mathbf{w}^{(i)} = (\mathbf{w}_{(f(i)+1)}, \mathbf{w}_{(f(i)+2)}, \ldots, \mathbf{w}_{jMax})$ be the resulting weight vector, and let $c_{(f(i)+1)}$ be the directly estimated cost of upgrading predictor variable 'i' from level $f_{(i)}$ to level $(f_{(i)} + 1)$. Then, for each level $p \in [(f_{(i)} + 2), j_{Max}]$, the corresponding upgrade cost would be estimated as

$$c_p = c_{(f(i)+1)} \times \left(\mathbf{w}_p / \mathbf{w}_{(f(i)+1)} \right).$$

Estimation of value can be done in a similar manner. Let a_{pq} be the numeric ratio that represents the benefit value of level 'p' of the dependent variable relative to the corresponding benefit value of level 'q.' Given h_{Cur} the current level of the dependent variable, PC data a_{pq} would be collected for each integer pair (p, q) where $p \in [(h_{Cur} + 1), h_{max}]$ and $q \in [h_{Cur}, (h_{max} - 1)]$. Let $\mathbf{w}^{(o)} = (\mathbf{w}_{hCur}, \mathbf{w}_{(hCur+1)}, \ldots, \mathbf{w}_{hMax})$ be the resulting weight vector, and let v_{hMin} be the directly estimated benefit value of the dependent variable at its current level. Then, for each level $p \in [(h_{Cur} + 1), h_{max}]$, the corresponding benefit value would be estimated as $v_p = v_{hCur} \times (\mathbf{w}_p / \mathbf{w}_{hCur})$.

7.4.3 Description of the Process

Step 1: Assess the state of the organization in terms of the levels of all predictor and dependent variables.

Step 2: Use a PC-based process to determine the Upgrade Costs.

Step 3: Use a PC-based process to determine the Benefit Values.

Step 4: Formulate and solve the corresponding Mathematical Programming Problem, and if necessary do sensitivity analysis.

7.5 ILLUSTRATIVE EXAMPLE

Our example involves three statistically significant predictors: *System Quality* (SQ), *Ease-of-Use* (EU), and *Utilization* (UT) with the corresponding coefficients of *0.507, 0.131*, and *0.349*, respectively. The intercept is 0.303, and so it follows that the regression equation is

$$Y = 0.303 + 0.507 \times SQ + 0.131 \times EU + 0.349 \times UT.$$

If we assume that the three predictor variables are at level 4, then the Y = 4.33 and its associated discrete level is 4 since Round (4.33) equals 4.

Let the budgeted amount be $B = 8.00$, and the Upgrade Costs and Benefit Values be displayed in Tables 7.1 and 7.2, respectively. Given this input data Table 7.3 presents the non-binary constraints and Table 7.4 displays the resulting solution levels for the three predictor variables and the output variable. These results indicate that the total benefit value is 16 while the total upgrade cost is 6.70. It should be noted that the actual gain in benefit is 10 (i.e. 16 – 6) so an investment of 6.70 units resulted in a gain in benefit of 10 units. If the net gain in benefit was less than the total upgrade costs then this would indicate that focus should not be on upgrading the levels of the input variables but rather on improving the process by which inputs produce outputs.

TABLE 7.1

Upgrade Costs

	Upgrade Costs		
Predictor	**4–5**	**4–6**	**4–7**
SQ	1.85	2.50	4.35
EU	1.55	1.75	3.25
UT	1.95	2.35	3.85

TABLE 7.2

Benefit Values

Level	4	5	6	7
Value	6.00	10.00	16.00	18.00

TABLE 7.3

Constraints and Objective Function

Constraints	Description
1a	SQ4 + SQ5 + SQ6 + SQ7 = 1
	EU4 + EU5 + EU6 + EU7 = 1
	UT4 + UT5 + UT6 + UT7 = 1
1c	4SQ4 + 5SQ5 + 6SQ6 + 7SQ7 − ZSQ = 0
	4EU4 + 5EU5 + 6EU6 + 7EU7 − ZEU = 0
	4UT4 + 5UT5 + 6UT6 + 7UT7 − ZUT = 0
1d	0.00SQ4 + 1.85SQ5 + 2.50SQ6 + 4.35SQ7 − BDSQ = 0
	0.00EU4 + 1.55EU5 + 1.75EU6 + 3.25EU7 − BDEU = 0
	0.00UT4 + 1.95UT5 + 2.35UT6 + 3.85UT7 − BDUT = 0
1e	BDSQ + BDEU + BDUT − BCOST = 0
	BCOST ≤ 8

Objective Function	Max 0.303 + 0.507ZSQ + 0.131ZEU + 0.349ZUT

TABLE 7.4

Solution

Predictor	Level	Cost
SQ	7	4.35
EU	4	0.00
UT	6	2.35
	Total	6.70

Dependent	Level	Value
Y_{Max}	6.47	
Y_{MaxRnd}	6 = Round(6.47)	16.00

7.6 DISCUSSION

In this section, we discuss some concerns that may arise as to the generality of the proposed method given the material presented in Section 7.4.

7.6.1 Accommodating Mediator and Multiple Dependent Variables

The mathematical programming formulation that was presented in Section 7.4.1 may appear to be appropriate only in the situation where there is no mediator variable or if there is only a single dependent variable. However, constraints similar to 1a– 1e can be formulated to represent the relationships of each mediator variable and its statistically significant antecedents, and each dependent variable and its statistically significant antecedents.

If multiple dependent variables are present, the objective function could be based on summing the corresponding individual benefit value functions. Alternately, it could be addressed as a multi-objective programming problem.

7.6.2 Estimating Upgrade Costs and Benefit Values

The approach that was presented in Section 7.4.2 may appear to be appropriate only in some situations, with other situations seeming to require that multiple criteria need to be considered in determining the upgrade costs, and similarly for the benefit values. In such situations, the AHP of Saaty's (1980) approach could be used where the top levels of the hierarchy correspond to the hierarchical structuring of the set of criteria and the lowest level would be for the alternatives. For the case of the upgrade costs for a given predictor variable, its relevant levels would correspond to the alternatives in this AHP framework. For the case of the benefit values, the relevant levels of the output variable would correspond to the alternatives. PC elicitation would be used to determine the relative weights of the criteria, and also the relative weights of each alternative with respect to each leaf-level criterion. This approach has been used in multiple real-world applications of the AHP in organizational settings. It should also be noted that there are various commercial level AHP software for individual and group decision-making that should make the application of this step convenient to the decision-maker.

7.7 CONCLUSION

In this chapter, we have presented and illustrated a hybrid mathematical programming/ MCDA-based approach for making actionable the results of quantitative behavioural science research projects whose outputs are causal theoretical models that both explain and predict (Gregor, 2006), and in which the items associated with the constructs were measured on a Likert scale. Based on the fact that MCDA methods such as the AHP have been successfully applied in real-world organizational settings, we have good reason to believe that the proposed approach is feasible in such settings. While there are still opportunities for improving this work, we would argue that it offers a path to making research results actionable for studies that have causal models with high explanatory and predictive powers.

REFERENCES

An, S.-H., Kim, G.-H., & Kang, K.-I. (2007). A case-based reasoning cost estimating model using experience by analytic hierarchy process. *Building and Environment, 42*(7), 2573–2579. https://doi.org/10.1016/j.buildenv.2006.06.007

Bryson, N. (1995). A goal programming method for generating priority vectors. *Journal of the Operational Research Society, 46*(5), 641–648. https://doi.org/10.1057/jors.1995.88

Bryson, N. (1996). Group decision-making and the analytic hierarchy process: Exploring the consensus-relevant information content. *Computers & Operations Research, 23*(1), 27–35. https://doi.org/10.1016/0305-0548(96)00002-H

Bryson, N., & Joseph, A. (1999). Generating consensus priority point vectors: A logarithmic goal programming approach. *Computers & Operations Research, 26*(6), 637–643. https://doi.org/10.1016/S0305-0548(98)00083-5

Ghasemaghaei, M., Hassanein, K., & Turel, O. (2016). Increasing firm agility through the use of data analytics: The role of fit. *Decision Support Systems, 101*, 95–105. https://doi.org/10.1016/j.dss.2017.06.004

Gregor, S. (2006). The nature of theory in information systems. *MIS Quarterly, 30*(3), 611–642.

Gulati, R. (2007). Tent poles, tribalism, and boundary spanning: The rigor-relevance debate in management research. *Academy of Management Journal, 50*(4), 775–782. https://doi.org/10.5465/amj.2007.26279170

Pavlou, P. A. (2003). Consumer acceptance of electronic commerce: Integrating trust and risk with the technology acceptance model. *International Journal of Electronic Commerce, 7*(3), 101–134. https://doi.org/10.1080/10864415.2003.11044275

Saaty, T. (1980). *The analytic hierarchy process: Planning, priority setting, resource allocation.* New York: McGraw Hill.

Appendix A
Binary Integer Programming Problem – Select the Candidate Models (SCM)

Given a set of candidate models, the research team may be interested in determining what would be a promising subset of these models for which data can be collected on a questionnaire that will result in at least the required N_{Min} usable observations via sampling the relevant target population.

Let:

- J be the index set of the models included in M;
- I be the index set of all variables included in the candidate models in M and I_j be the index set of the variables included in model m_j;
- b_i be the number of questionnaire items associated with variable I;
- b_{MaxTot} be an estimate of the maximum total number of question-naire items that will result in N_{Min} usable observations;
- v_j be the research team's estimate of the relative value of the con-tributions offered by model m_j, where $v_j \in (0, 1)$ with larger values indicating greater value;
- x_i be a binary decision variable that indicates whether all items for model variable i would be included in the questionnaire;
- z_j be a binary decision variable that indicates whether model m_j would be fully covered by the questionnaire.

The subset of the models that should be covered by the questionnaire could be obtained by solving the following binary integer programming problem **SCM**:

$$Max \sum_{j \in J} v_j z_j$$

s.t.

$$(1) \quad |I_j| z_j - \sum_{i \in I_j} x_i \leq 0 \qquad \forall j \in J,$$

$$(2) \quad \sum_{i \in I} b_i x_i \leq b_{MaxTot},$$

(3) $\qquad y_j \in \{0, 1\} \qquad\qquad \forall j \in J,$

(4) $\qquad x_i \in \{0, 1\} \qquad\qquad \forall i \in I.$

Constraint 1 ensures that z_j can be 1 only if all of the variables of model m_j are covered by the questionnaire. Constraint 2 ensures that the total number of items on the questionnaire is not greater than the acceptable level of b_{Max}. The objective function (i.e., $\sum_{j \in J} v_j z_j$) reflects the total estimated value of questionnaire offering full coverage for the selected subset of models in M.

Appendix B

TABLE B.1

Constructs, Measurement Items, Descriptive Statistics, and Sources

Construct	Code	Item	Weight	Loading (*t*-value)	Source(s)
IS Security Compliance Intention[a]	SINT1	It is my intention to continue to comply with the organisation's IS security policy/ guidelines.	0.291***	0.777 (8.3866)***	Ifinedo (2014) Ifinedo (2012)
	SINT2	I am likely to follow the organisation's IS security policy/ guidelines in the future.	0.431***	0.948 (67.8025)***	
	SINT3	I would follow the organisation's security policy/guidelines whenever possible.	0.403***	0.909 (28.3522)***	
IS Security Compliance Attitude[a]	SATT1	To me, complying with organisational IS security policy/ guidelines is a necessity.	0.383***	0.965 (100.5934)***	Ifinedo (2014) Ifinedo (2012)
	SATT2	To me, complying with organisational IS security policy/ guidelines is beneficial.	0.368***	0.945 (62.5966)***	
	SATT3	To me, complying with organisational IS security policies/ guidelines is a good idea.	0.322***	0.880 (15.9301)***	

(*Continued*)

TABLE B.1 (Continued)

Construct	Code	Item	Weight	Loading (*t*-value)	Source(s)
IS Security Self-efficacy[a]	SELF1	I have the necessary skills to protect myself from information security violations.	0.411***	0.896 (36.4327)***	Ifinedo (2014) Ifinedo (2012)
	SELF2	I have the expertise to implement preventative measures to stop people from getting my confidential work information.	0.271***	0.867 (21.0143)***	
	SELF3	I have the skills to implement preventative measures to stop people from damaging my work computer.	0.444***	0.894 (30.1064)***	
General IS Security Awareness[a]	GSAW1	Overall, I am aware of potential information/cybersecurity threats and their negative consequences.	0.335***	0.780 (15.2744)***	Donalds and Osei-Bryson (2020); Donalds and Osei-Bryson (2017) Donalds (2015); Bulgurcu, Cavusoglu, and Benbasat (2010)
	GSAW2	I understand the concerns regarding IS security threats and the risks they pose in general.	0.441***	0.910 (50.1324)***	
	GSAW3	I have sufficient knowledge about the cost of potential IS security incidents.	0.377***	0.895 (35.5251)***	
General IS Security Orientation[a]	GSOR1	I usually read information security bulletins, newsletters, and/or messages.	0.4378***	0.891 (33.8621)***	Donalds and Osei-Bryson (2020); Donalds and Osei-Bryson (2017); Ng, Kankanhalli, and Xu (2009)
	GSOR2	I am usually concerned about information security incidents and try to take actions to prevent them.	0.393***	0.904 (39.0012)***	
	GSOR3	I am usually mindful about computer security.	0.310***	0.823 (18.9043)***	

(Continued)

TABLE B.1 (Continued)

Construct	Code	Item	Weight	Loading (*t*-value)	Source(s)
Response Cost[a]	COST1	To me, too much effort is required to follow the rules outlined in the IS security policy/ guidelines.	0.161	0.682[c] (4.1949)***	Boss, Galletta, Lowry, Moody, and Polak (2015); Ifinedo (2012); Milne, Orbell, and Sheeran (2002) questions are adapted to this context
	COST2	To me, the effort required to follow the IS security policy/ guidelines outweighs the benefits.	0.536**	0.924 (8.2008)***	
	COST3	To me, following ALL the rules in the IS security policy/ guidelines would require considerable effort.	0.439**	0.900 (8.5072)***	
Personal Norm[a]	PNRM1	To me, it is unacceptable to not follow ALL rules and measures outlined in the organisation's IS security policy/ guidelines.	0.510***	0.893 (27.9212)***	Ifinedo (2014) (H. Li, Zhang, & Sarathy, 2010)
	PNRM2	It is not a trivial offense if I do not follow the organisation's IS security policy/ guidelines.	0.139*	0.487[c] (35.6462)***	
	PNRM3	To me, it is unacceptable to ignore my organisation's IS security policy/ guidelines.	0.528***	0.904 (5.0829)***	

(*Continued*)

TABLE B.1 (Continued)

Construct	Code	Item	Weight	Loading (*t*-value)	Source(s)
Normative Belief[a] (Subjective Norm)	BELF1	Top management thinks that I should follow organisational IS security policy/ guidelines.	0.546***	0.899 (32.3475)***	Siponen, Mahmood, and Pahnila (2014); Bulgurcu et al. (2010);Herath and Rao (2009)
	BELF2	My boss thinks that I should follow organisational IS security policy/guidelines.	0.563***	0.905 (26.9153)***	
	BELF3	My colleagues think that I should follow organisational IS security policy/guidelines.	1.000	1.000	
	BELF4	The information security department in my organisation thinks that I should follow organisational IS security policy/ guidelines.	0.557***	0.929 (53.5984)***	
	BELF5	Computer technical specialists in my organisation think that I should follow organisational IS security policy/ guidelines.	0.525***	0.920 (28.4178)***	
Detection Certainty[a]	DETC1	To me, employee computer practices are properly monitored for policy/guideline violations.	0.413***	0.912 (46.9468)***	Herath and Rao (2009); third question is self-developed
	DETC2	If I violate organisational IS security policy/ guidelines, I would probably be caught.	0.431***	0.943 (79.4874)***	
	DETC3	To me, the likelihood that the organisation would discover that an employee failed to follow the rules in the IS security policies/ guidelines is likely.	0.267***	0.811 (15.7914)***	

(*Continued*)

TABLE B.1 (Continued)

Construct	Code	Item	Weight	Loading (*t*-value)	Source(s)
Punishment Severity[a]	PUNS1	The organisation disciplines employees who break IS security policy/guidelines.	0.416***	0.899 (46.3708)***	Herath and Rao (2009)
	PUNS2	My organisation would terminate or terminates employees who repeatedly break IS security policy/ guidelines.	0.371***	0.906 (42.2930)***	
	PUNS3	If I were caught violating organisational IS security policy/ guidelines, I would be severely punished.	0.352***	0.824 (18.1173)***	
Response Efficacy[a]	RESP1	To me, following the IS security policy/ guidelines is a good way to reduce information security threats.	0.370***	0.859 (23.2243)***	L. Li et al. (2019); Siponen et al. (2014)
	RESP2	To me, following ALL the rules outlined in the IS security policy/guidelines is a good way to protect organisational information systems.	0.351***	0.916 (32.2644)***	
	RESP3	If I were to follow ALL the rules in the IS security policy/ guidelines, I would lessen my chances of experiencing information security incidents.	0.387***	0.932 (51.6522)***	

(*Continued*)

TABLE B.1 (Continued)

Construct	Code	Item	Weight	Loading (*t*-value)	Source(s)
Marker Variable[b]	COMB1	Please rate how quickly your IT department is able to detect changes in customer satisfaction.	N/A	N/A	Balozian, Leidner, and Warkentin (2017)
	COMB2	Please assess the extent to which your IT department contributes to the organisation process improvement.	N/A	N/A	
	COMB3	Please assess the extent to which your IT department contributes to the organisation operating efficiency.	N/A	N/A	

Notes:
[a] identifies construct's items scale: 5 pt. Likert from "strongly disagree" to "strongly agree."
[b] identifies construct's items scale: 7 pt. Likert from "no extent" to "very great extent."
[c] identifies indicator with low loading, which is subsequently dropped from the model.
* $p < 0.05$,
** $p < 0.01$,
*** $p < 0.001$.
N/A = not applicable.

REFERENCES

Balozian, P., Leidner, D., & Warkentin, M. (2017). Managers' and employees' differing responses to security approaches. *Journal of Computer Information Systems*, *59*(3), 197–210. http://doi.org/10.1080/08874417.2017.1318687

Boss, S. R., Galletta, D. F., Lowry, P. B., Moody, G. D., & Polak, P. (2015). What do systems users have to fear? Using fear appeals to engender threats and fear that motivate protective security behaviors. *MIS Quarterly*, *39*(4), 837–864.

Bulgurcu, B., Cavusoglu, H., & Benbasat, I. (2010). Information security policy compliance: An empirical study of rationality-based beliefs and information security awareness. *MIS Quarterly*, *34*(3), 523–548.

Donalds, C. (2015). *Cybersecurity policy compliance: An empirical study of Jamaican government agencies*. Paper presented at *the SIG GlobDev Pre-ECIS Workshop*, Munster, Germany.

Donalds, C., & Osei-Bryson, K.-M. (2017). *Exploring the Impacts of Individual Styles on Security Compliance Behavior: A Preliminary Analysis*. Paper presented at the *SIG ICT in Global Development, 10th Annual Pre-ICIS Workshop*, Seoul, Korea.

Donalds, C., & Osei-Bryson, K.-M. (2020). Cybersecurity compliance behavior: Exploring the influences of individual decision style and other antecedents. *International Journal of Information Management, 51.* https://doi.org/10.1016/j.ijinfomgt.2019.102056

Herath, T., & Rao, H. R. (2009). Protection motivation and deterrence: A framework for security policy compliance in organisations. *European Journal of Information Systems, 18*(2), 106–125. https://doi.org/10.1057/ejis.2009.6

Ifinedo, P. (2012). Understanding information systems security policy compliance: An integration of the theory of planned behavior and the protection motivation theory. *Computers & Security, 31*(1), 83–95. https://doi.org/10.1016/j.cose.2011.10.007

Ifinedo, P. (2014). Information systems security policy compliance: An empirical study of the effects of socialisation, influence, and cognition. *Information & Management, 51*(1), 69–79.

Li, H., Zhang, J., & Sarathy, R. (2010). Understanding compliance with internet use policy from the perspective of rational choice theory. *Decision Support Systems, 48*(4), 635–645. https://doi.org/10.1016/j.dss.2009.12.005

Li, L., He, W., Xu, L., Ash, I., Anwar, M., & Yuan, X. (2019). Investigating the impact of cybersecurity policy awareness on employees' cybersecurity behavior. *International Journal of Information Management, 45*, 13–24. https://doi.org/10.1016/j.ijinfomgt.2018.10.017

Milne, S., Orbell, S., & Sheeran, P. (2002). Combining motivational and volitional interventions to promote exercise participation: Protection motivation theory and implementation intentions. *British Journal of Health Psychology, 7*(2), 163–184. http://doi.org/10.1348/135910702169420

Ng, B.-Y., Kankanhalli, A., & Xu, Y. C. (2009). Studying users' computer security behavior: A health belief perspective. *Decision Support Systems, 46*(4), 815–825.

Siponen, M. T., Mahmood, M. A., & Pahnila, S. (2014). Employees' adherence to information security policies: An exploratory field study. *Information & Management, 51*(2), 217–224. http://doi.org/10.1016/j.im.2013.08.006

Appendix C

TABLE C.1

Item Loading and Cross Loadings

Indicator	PNRM	GSOR	SELF	GSAW	RESP	DETC	SATT	COST	SINT	PUNS	MGMT	COLL	IT-D
PNRM1	**0.915**	0.466	0.259	0.434	0.504	0.303	0.554	-0.149	0.531	0.337	0.430	0.329	0.378
PNRM3	**0.896**	0.445	0.187	0.386	0.514	0.425	0.496	-0.218	0.492	0.431	0.459	0.427	0.409
GSOR1	0.497	**0.891**	0.479	0.613	0.574	0.412	0.599	-0.207	0.603	0.281	0.553	0.436	0.465
GSOR2	0.492	**0.904**	0.506	0.626	0.546	0.446	0.549	-0.236	0.531	0.379	0.593	0.606	0.546
GSOR3	0.297	**0.823**	0.399	0.570	0.459	0.352	0.432	-0.256	0.418	0.390	0.429	0.471	0.376
SELF1	0.215	0.439	**0.885**	0.372	0.253	0.141	0.342	-0.007	0.303	0.203	0.277	0.324	0.315
SELF2	0.170	0.476	**0.883**	0.426	0.214	0.142	0.219	-0.033	0.208	0.233	0.267	0.292	0.266
SELF3	0.262	0.498	**0.894**	0.496	0.305	0.249	0.364	-0.104	0.334	0.212	0.377	0.370	0.339
GSAW1	0.327	0.571	0.493	**0.795**	0.418	0.213	0.450	-0.208	0.498	0.252	0.475	0.378	0.399
GSAW2	0.409	0.563	0.337	**0.899**	0.641	0.262	0.645	-0.227	0.601	0.251	0.543	0.421	0.461
GSAW3	0.436	0.658	0.453	**0.894**	0.588	0.251	0.551	-0.166	0.514	0.231	0.547	0.387	0.399
RESP1	0.460	0.554	0.296	0.546	**0.859**	0.339	0.685	-0.349	0.690	0.230	0.469	0.499	0.409
RESP2	0.506	0.523	0.239	0.580	**0.916**	0.402	0.706	-0.169	0.655	0.278	0.537	0.490	0.469
RESP3	0.552	0.567	0.263	0.606	**0.932**	0.454	0.782	-0.191	0.722	0.332	0.598	0.458	0.527
DETC1	0.384	0.386	0.212	0.263	0.438	**0.909**	0.441	-0.072	0.344	0.468	0.383	0.361	0.403
DETC2	0.404	0.458	0.182	0.291	0.435	**0.945**	0.410	-0.023	0.374	0.501	0.338	0.369	0.349
DETC3	0.248	0.401	0.153	0.178	0.279	**0.815**	0.300	0.043	0.236	0.531	0.292	0.357	0.297
SATT1	0.587	0.602	0.349	0.617	0.807	0.432	**0.965**	-0.228	0.852	0.330	0.671	0.482	0.559
SATT2	0.550	0.592	0.371	0.578	0.802	0.456	**0.945**	-0.239	0.811	0.313	0.650	0.486	0.569
SATT3	0.478	0.509	0.266	0.593	0.621	0.329	**0.881**	-0.166	0.682	0.267	0.575	0.427	0.503
COST2	-0.196	-0.273	-0.030	-0.207	-0.273	-0.042	-0.228	**0.944**	-0.255	-0.043	-0.230	-0.230	-0.136

(Continued)

TABLE C.1 (Continued)

Indicator	PNRM	GSOR	SELF	GSAW	RESP	DETC	SATT	COST	SINT	PUNS	MGMT	COLL	IT-D
COST3	-0.170	-0.201	-0.088	-0.224	-0.199	-0.001	-0.187	**0.893**	-0.188	0.002	-0.211	-0.173	-0.125
SINT1	0.347	0.428	0.343	0.454	0.474	0.280	0.631	-0.155	**0.776**	0.203	0.472	0.314	0.343
SINT2	0.598	0.586	0.287	0.599	0.764	0.369	0.833	-0.246	**0.948**	0.344	0.700	0.485	0.576
SINT3	0.514	0.559	0.251	0.581	0.740	0.312	0.749	-0.234	**0.909**	0.299	0.685	0.473	0.554
PUNS1	0.384	0.353	0.276	0.256	0.305	0.504	0.309	-0.032	0.328	**0.896**	0.286	0.312	0.349
PUNS2	0.380	0.267	0.143	0.205	0.239	0.424	0.263	-0.015	0.256	**0.893**	0.292	0.280	0.341
PUNS3	0.344	0.400	0.202	0.277	0.267	0.511	0.282	-0.019	0.269	**0.840**	0.315	0.460	0.409
BELF1	0.434	0.600	0.284	0.546	0.519	0.416	0.625	-0.243	0.635	0.366	**0.899**	0.570	0.656
BELF2	0.450	0.498	0.351	0.545	0.550	0.273	0.602	-0.191	0.654	0.249	**0.905**	0.425	0.635
BELF3	0.415	0.575	0.375	0.458	0.534	0.404	0.501	-0.223	0.491	0.403	0.550	**1.000**	0.680
BELF4	0.366	0.518	0.355	0.443	0.428	0.308	0.524	-0.136	0.543	0.346	0.702	0.684	**0.929**
BELF5	0.438	0.469	0.289	0.458	0.538	0.427	0.559	-0.127	0.512	0.432	0.618	0.570	**0.920**

Note: PNRM = Personal Norms; GSOR = General Security Orientation; SELF = Security Self-Efficacy; GSAW = General Security Awareness; RESP = Efficacy; DETC = Detection Certainty; SATT = Security Attitude; COST = Response Cost; SINT = Security Intention; PUNS = Punishment Severity; MGMT = Normative Belief – Management; COLL = Normative Belief – Colleague; IT-D = Normative Belief – IT Department.

Appendix D

TABLE D.1

Research Hypotheses, Results of Their Evaluations in This and Previous Studies

Hypothesis		Any Previous Study		This Research
		Supported	Rejected	Supported
H1:	IS security attitude → ISP compliance intention	Bauer and Bernroider (2017); Bulgurcu, Cavusoglu, and Benbasat (2010); D'Arcy and Lowry (2019); Hu, Dinev, Hart, and Cooke (2012); Ifinedo (2012); Ifinedo (2014); Siponen, Mahmood, and Pahnila (2014)	Aurigemma and Mattson (2017); Herath and Rao (2009)	Yes
H2a:	IS security self-efficacy → ISP compliance intention	Bulgurcu et al. (2010); Chan, Woon, and Kankanhalli (2005); D'Arcy and Lowry (2019); Herath and Rao (2009); Ifinedo (2012); Ifinedo (2014); Donalds and Osei-Bryson (2020); Lee and Larsen (2009); Ng, Kankanhalli, and Xu (2009); Siponen et al. (2014);	Wall, Palvia, and Lowry (2013)	No
H2b:	IS security self-efficacy → IS security attitude	Herath and Rao (2009)		No
H3a:	Norm. Belief (Management) → ISP compliance intention	Bulgurcu et al. (2010)[a]; Herath and Rao (2009)[b]; Ifinedo (2014)[a]; Siponen et al. (2014)		Yes
H3b:	Norm. Belief (IT-Dept.) → ISP compliance intention			No

(*Continued*)

TABLE D.1 (Continued)

Hypothesis		Any Previous Study		This Research
		Supported	Rejected	Supported
H3c:	Norm. Belief (Colleague) → ISP compliance intention			No
H4:	**Response efficacy → ISP compliance intention**	Ifinedo (2012); Lee and Larsen (2009); Wall et al. (2013)	Siponen et al. (2014);	**Yes**
H5:	Response cost → ISP compliance intention	Lee and Larsen (2009)	Ifinedo (2012);	No
H6:	Punishment severity → ISP compliance intention	D'Arcy, Hovav, and Galletta (2009); Herath and Rao (2009)	Balozian, Leidner, and Warkentin (2017); Li, Sarathy, Zhang, and Luo (2014);	No
H7:	Detection certainty → ISP compliance intention	Herath and Rao (2009); Li et al. (2014)	Balozian et al. (2017); D'Arcy et al. (2009)	No
H8a:	**General security awareness → IS security attitude**	Bauer and Bernroider (2017); Bulgurcu et al. (2010); Safa et al. (2015)		**Yes**
H8b:	General security awareness → ISP compliance intention	Donalds and Osei-Bryson (2020); Koohang, Nowak, Paliszkiewicz, and Horn Nord (2020)		No
H8c:	**General security awareness → IS security self-efficacy**			**Yes**
H8d:	**General security awareness → Punishment severity**	D'Arcy et al. (2009)		**Yes**
H8e:	**General security awareness → Detection certainty**	D'Arcy et al. (2009)		**Yes**
H9a:	General security orientation → ISP compliance intention	Donalds and Osei-Bryson (2020)	Ng et al. (2009)	No

(Continued)

TABLE D.1 (Continued)

Hypothesis		Any Previous Study		This Research
		Supported	Rejected	Supported
H9b:	General security orientation → IS security attitude			Yes
H10a:	Personal norms → IS security attitude	Ifinedo (2014); Safa, von Solms, and Furnell (2016)		Yes
H10b:	Personal norms → ISP compliance intention	Li et al. (2014) Yazdanmehr and Wang (2016)		No

[a] Normative belief was modelled as a multi-item reflective construct.
[b] Normative belief was modelled as a formative construct.

REFERENCES

Aurigemma, S., & Mattson, T. (2017). Privilege or procedure: Evaluating the effect of employee status on intent to comply with socially interactive information security threats and controls. *Computers & Security, 66*, 218–234.

Balozian, P., Leidner, D., & Warkentin, M. (2017). Managers' and employees' differing responses to security approaches. *Journal of Computer Information Systems, 59*(3), 197–210. doi:10.1080/08874417.2017.1318687

Bauer, S., & Bernroider, E. W. N. (2017). From information security awareness to reasoned compliant action: Analyzing information security policy compliance in a large banking organization. *The DATA BASE for Advances in Information Systems, 48*(3), 44–68.

Bulgurcu, B., Cavusoglu, H., & Benbasat, I. (2010). Information security policy compliance: An empirical study of rationality-based beliefs and information security awareness. *MIS Quarterly, 34*(3), 523–548.

Chan, M., Woon, I., & Kankanhalli, A. (2005). Perceptions of information security in the workplace: Linking information security climate to compliant behavior. *Journal of Information Privacy & Security, 1*(3), 18–41.

D'Arcy, J., & Lowry, P. B. (2019). Cognitive-affective drivers of employees' daily compliance with information security policies: A multilevel, longitudinal study. *Information Systems Journal, 29*(1), 43–69. https://doi.org/10.1111/isj.12173

D'Arcy, J., Hovav, A., & Galletta, D. (2009). User awareness of security countermeasures and its impact on information systems misuse: A deterrence approach. *Information Systems Research, 20*(1), 79–98.

Donalds, C., & Osei-Bryson, K.-M. (2020). Cybersecurity compliance behavior: Exploring the influences of individual decision style and other antecedents. *International Journal of Information Management, 51*. https://doi.org/10.1016/j.ijinfomgt.2019.102056

Herath, T., & Rao, H. R. (2009). Protection Motivation and Deterrence: A Framework for Security Policy Compliance in Organisations. *European Journal of Information Systems, 18*(2), 106–125. https://doi.org/10.1057/ejis.2009.6

Hu, Q., Dinev, T., Hart, P., & Cooke, D. (2012). Managing employee compliance with information security policies: The critical role of top management and organizational culture. *Decision Sciences, 43*(4), 615–659.

Ifinedo, P. (2012). Understanding information systems security policy compliance: An integration of the theory of planned behavior and the protection motivation theory. *Computers & Security, 31*(1), 83–95. https://doi.org/10.1016/j.cose.2011.10.007

Ifinedo, P. (2014). Information systems security policy compliance: An empirical study of the effects of socialisation, influence, and cognition. *Information & Management, 51*(1), 69–79.

Koohang, A., Nowak, A., Paliszkiewicz, J., & Horn Nord, J. (2020). Information security policy compliance: Leadership, trust, role values, and awareness. *Journal of Computer Information Systems, 60*(1), 1–8. https://doi.org/10.1080/08874417.2019.1668738

Lee, Y., & Larsen, K. R. (2009). Threat or coping appraisal: Determinants of SMB executives' decision to adopt anti-malware software. *European Journal of Information Systems, 18*(2), 177–187. http://doi.org/10.1057/ejis.2009.11

Li, H., Sarathy, R., Zhang, J., & Luo, X. (2014). Exploring the effects of organizational justice, personal ethics and sanction on internet use policy compliance. *Information Systems Journal, 24*(6), 479–502. https://doi.org/10.1111/isj.12037

Ng, B.-Y., Kankanhalli, A., & Xu, Y. C. (2009). Studying Users' Computer Security Behavior: A Health Belief Perspective. *Decision Support Systems, 46*(4), 815–825.

Safa, N. S., Sookhak, M., Von Solms, R., Furnell, S., Ghani, N. A., & Herawan, T. (2015). Information security conscious care behaviour formation in organizations. *Compters & Security, 53*, 65–78. http://doi.org/10.1016/j.cose.2015.05.012

Safa, N. S., von Solms, R., & Furnell, S. (2016). Information security policy compliance model in organizations. *Computers & Security, 56*(C), 70–82.

Siponen, M. T., Mahmood, M. A., & Pahnila, S. (2014). Employees' adherence to information security policies: An exploratory field study. *Information & Management, 51*(2), 217–224. http://doi.org/10.1016/j.im.2013.08.006

Wall, J. D., Palvia, P., & Lowry, P. B. (2013). Control-related motivations and information security policy compliance: The role of autonomy and efficacy. *Journal of Information Privacy and Security, 9*(4), 52–79. https://doi.org/10.1080/15536548.2013.10845690

Yazdanmehr, A., & Wang, J. (2016). Employees' information security policy compliance: A norm activation perspective. *Decision Support Systems, 92*, 36–46.

Index

Note: Pages in *italics* refer to figures and pages in **bold** refer to tables.

Printed in the United States
by Baker & Taylor Publisher Services